WALK THROUGH THE WILDERNESS

WALK THROUGH THE WILDERNESS

DON RICHARDS • CLIVE WALKER

PURNELL
CAPE TOWN • JOHANNESBURG • LONDON

PUBLISHED BY PURNELL & SONS (PTY.) LTD.
70 KEEROM STREET, CAPE TOWN

SBN 360 00275 7

TEXT SET IN 11 ON 13 PT BASKERVILLE, PRINTED AND BOUND
BY RUSTICA PRESS (PTY.) LTD., WYNBERG, CAPE, SOUTH AFRICA

DEDICATED

TO BARBARA AND CONITA

Preface

In recent years, biologists have realized that no plant or animal can be truly understood when considered in isolation. Every living thing forms part of a community, and can be understood only within this context. Man is only one species among a multitude with which he is inextricably involved.

Walk through the Wilderness is a study of the inter-relationships of all living things which exist in all natural communities, including man. The real value of this book is that it has been written for South Africans.

If man as a species is to live harmoniously with nature, then it is important that he understands the system.

Walk through the Wilderness is neither scientific nor complicated. It is an informative account of how a South African can appreciate and become involved in his environment. It shows that conservation of the South African environment is not taboo—it is something which every individual is entitled to know.

The book is the interpretation of Don Richards and Clive Walker, whose years of environmental experience and research are condensed in its pages.

Clive Walker is responsible for the illustrations and many photographs which highlight the chapters, making it possible for every South African to know, understand and enjoy his environment.

Foreword

by

IAN PLAYER

I have had the privilege of knowing and working with the authors of this book. Both have had a long association with the Wilderness Leadership School and have contributed much to its continuing success in the field of wilderness and environmental education.

Don Richards and Clive Walker are practical men with wide experience in the field of natural history. Clive Walker's reputation as a wildlife artist has spread to the United States, where his work is commanding justifiably high prices.

Don Richards, now the leader of 'Joint Venture', a combined exercise of the Wilderness Leadership School and the Wildlife Society, has for years been the best teacher in his field. Both these men are dedicated in the true sense of the word and have become prominent in the struggle to educate people of all races to the value of our natural areas.

The book they have written is important in that it can be read and appreciated by people of all ages. It is a sound mixture of science, art and general knowledge and the presentation of all the material is outstanding. The information it contains covers a wide field: forests, water, grasslands, energy flow, oceans, estuaries, desert and the treatment of snakebite.

I am very pleased to see that a section has been devoted to the Bushmen. In our atomic arrogance we are inclined to look down on the so-called primitive people, yet they were the only ones who lived in real harmony with the land and perhaps understood the meaning of happiness.

I have great pleasure in recommending this book and I only wish that something similar had been available to me when I started my career.

IAN PLAYER
Programme Director
INTERNATIONAL WILDERNESS LEADERSHIP FOUNDATION

Acknowledgements

FROM THE AUTHORS

The authors would like to express their gratitude to the many people who assisted in the compilation of *Walk through the Wilderness*. Special thanks are extended to:

The Natal Parks, Game and Fish Preservation Board
The Wilderness Leadership School
Mr Ian Player
Dr J. J. Brossy, for his contribution on the treatment on snake bite
Mr Gordon Bailey, senior ranger in charge Umfolozi Game Reserve—and his staff
Mr Gordon Forrest, senior lake ranger, Lake St Lucia
Mr Jim Feely, master in bushlore
Mrs Blyth Loutit, for botanical illustrations
Mrs J Clements
Mrs M. Aitken
Mr James Clarke, *Star*
Mr Dennis Cleaver
R. Patterson, Tvl. Snake Park
Museum Man of Science, information on Bushmen
University of Pretoria, Prof. J. du P. Bothma, Dr G. Theron
Miss S. B. Fox, who contributed greatly in the compilation of this book
Mr Alf Pexton, Mr B. Kemp and Simon Pillinger
Rustica Press
Frank Low and Jack Head, our publishers
To the game guards and trackers who have taught us so much:

It will be apparent that the compilation of this book has been the result of a team effort on the part of many contributors.

Wildlife Society of South Africa
Star, Argus Group
Mr Barry Clements
Mr Hugh Dent

All photographs (*black and white*) are by the author Clive Walker unless otherwise stated:

p. 21 Dennis Cleaver; p. 28 J. Pexton; p. 39 Alf Pexton; p. 42 Dennis Cleaver; p. 43 Art Sledricht; p. 68 Gordon Bailey; p. 69 B. Kemp; p. 72 B. Kemp; p. 75 G. Mills; p. 83 D. Richards; p. 84 D. Richards; p. 90 Purnell; p. 94 D. Richards; p. 95 D. Williams; p. 126 Arthur Boland; p. 128 D. Rushworth.

Contents

WALK THROUGH THE WILDERNESS . . .

An extract from the field diary of Clive Walker—13 June 1966:

At the confluence of the Shashi and the Tuli rivers there stands an ancient 'munye' tree. And set deep in the trunk is a marble tablet. The tablet bears an inscription which begins: 'As I walk through the wilderness of this world'.

When I first came upon the tablet, I was tracking a wounded elephant which subsequently died in the dry Tuli riverbed. Scores of vultures wheeling in a cloudless sky guided me to the east bank of the river where the elephant had fallen. As I approached, the vultures began to lumber forward in twos and threes, gathering momentum until they launched themselves skywards to join others circling high on head.

Standing alone among the mangled remains of a once majestic elephant, I noted how little time the bushveld scavengers had wasted. Elephant spoor in great numbers was apparent. One could avidly visualize their flapping ears and upraised trunks, sensing what had befallen one of their kind, and disappearing silently and quickly into the vast bushland.

Crossing to the west bank I discovered the tablet and fell to wondering who would have come to an area so remote to place a memorial. For this was indeed wilderness, untouched by the hand of Man. The forgotten pioneer ghosts of Fort Tuli were the only indications of Man's presence in that area. The inscription on the tablet continued: 'in memory of Arthur D. M. W. Vickers killed in the Galway Castle disaster 1918, erected by his brother.' I later established that the *Galway Castle* was torpedoed in the English Channel with a loss of 154 lives.

Vickers's brother Hariod erected the tablet in 1919. He had spent many a happy day in the Tuli wilderness with his partner Farrar. As early prospectors, Hariod Vickers and Farrar jointly pegged claims on what was formerly the Farvic Mine near Colleen Bawn some 40 miles [60 kilometres] north of Tuli. Through the years, the 'munye' tree had grown around the sides of the tablet, holding it firmly and protecting it from the elephant who might use the tree as a favourite rubbing-post.

The tablet, dedicated to Arthur Vickers, is well situated, for today only lion and elephant frequent the wilderness there.

WILDERNESS . . .

The term 'wilderness area' has been described by W. R. Bainbridge as follows:

'A wilderness area is any extensive undeveloped area, where the natural community of life is untrammelled and unhabited by man, who visits, but does not remain. It should retain an intrinsically wild appearance and character, or be capable of rehabilitation back into this wild state by relatively simple management measures. Its wild character must not be marred by more than minimal evidence of enduring man-made disturbance. The area must be well suited to a primitive or unconfined type of outdoor recreation and must provide the experience of isolation from the outside world, to those who enter. It will be aimed to provide high quality physical and spiritual recreation and inspiration, on a strictly limited basis, for the discriminating few who are prepared to enter by personal endeavour, in a spirit of acceptance of the Wilderness concept. Those who enter must depend on their own efforts and their own competence for survival. An essential attribute of a wilderness area is the provision of solitude, free from all mechanical disturbances.'

Preamble

'On foot, the pulse of Africa
comes through your boot'

Every item, large or small, in the massive kaleidoscope of flora and fauna upon this planet Earth fits into some logical niche within the wilderness scheme of things. Each unit finds nutrition, wards off enemies, survives seasonal changes and produces offspring. And, down through the centuries, the survivors are those which have adapted to changing circumstances—plant or animal, insect or reptile.

Few places on Earth offer better opportunities for the study of ecology than southern Africa. The infinite variety and frequent exclusivism of living things in this part of the world are truly astonishing and, as a consequence, utterly absorbing. Ecology—defined simply as the study of all living creatures in relation to their environment—involves a whole host of individual subjects and, in so doing, creates wonderful opportunities for exceptional botanical and biological observation anywhere; but here in this tip of the continent, such observation is sharpened and dramatized by the underlying presence of wildlife with its own particular brand of magic from prehistoric aeons.

Scanning an African mountain range, the untrained eye might well accept its monumental beauty as something so remote as to be seemingly lifeless. Only when given the chance of a closer look can one marvel first-hand at the incredible abundance of activity which starts at the foothills and continues up to the very peaks. Wide, sweeping grasses merge into closely-packed shrubs which grow into thick, natural forests —giving way finally to towering kranses and jagged pinnacles. Such rich and varied vegetation gives habitation and hospitality to a thousand species of beasts and birds of many shapes and sizes—all fitting into a clear, specific pattern of survival, planned millions of years ago and painstakingly adapted to countless upheavals of climate and geography.

Rocky outcrops amid pioneer grasses north of Capricorn are punctuated with great baobabs thrusting upwards like an ancient root system turned turtle. Low-lying areas where rivers flow in the rainy season often provide a perfect setting for groves of sinister fever trees. All around, one senses the primeval origins, mysteries and myths which made the wilderness of southern Africa one of mankind's very first homes.

Entirely free of charge is the thrilling spectacle of the haunting dawn of an African day—the swift sun of the subtropics rising, spinning and rejecting shadows across a characteristic vista of bushveld; or outlining the shape of a koppie in a wide expanse of platteland at its flattest.

As the day develops into mid-morning, the shy cheetah emerges from furtive scrub and lolls across a well-placed ant-heap to study antelope form, its sorrowful face markings and polka-dot coat well suited to immediate environment. The yellow beak of a hornbill, Africa's mini-toucan, flashes briefly in the dazzling sun, while never-still silhouettes of baboon seem ever distrustful of the hot, still, comparative quiet.

While a pair of reverend fish eagles perch observantly on a tall dead tree, nimble kingfishers dart in and out at lower levels. Constant communion between the countless feathered friends and enemies offers staccato contrast to the buzzing billions of insects. Shadowy figures with twitching ears glide to and fro among the prodigious variety of acacia and the tall, tawny grass.

Southern Africa is by no means all savannah, however.

Try following a lovely game trail through the Tsitsikama of the Knysna Great Forest—among noble trees that were there before the time of Christ. The same African midday sun penetrates those swaying branches and dangling epiphytes, illuminating momentarily a moss-covered log. Move it casually with a boot, and observe how the moist soil beneath it comes alive with a myriad of angry, scuttling beetles, woodborers, fishmoths and sundry creepy-crawlies.

Other senses can be exercised as one walks in the wilderness. The more perceptive might detect the rattling of termites among leaf litter, or inhale the aromatic charm of dense bush.

No less fascinating is it to lie upon the sands of a wild shore washed by the Indian Ocean, and to watch tiny crabs playing hide-and-seek with a gentle tide. Or to imagine the sounds of the sea in an abandoned shell. Or, for that matter, to gaze with wonderment into a clear Cape mountain stream where every stony conglomerate hides a handsome trout.

Systematically, and not so slowly, the development of your awareness begins to amaze and delight you. The call of a bird as yet unknown, the clear-cut spoor in dried mud of a big cat, the dainty hoof-prints of a tiny gazelle, or the peculiar formation of an aloe . . . all this and so very much more makes the personal discovery of wilderness a never-ending, completely rewarding process. Moreover, how forcibly are we reminded of our *very* insignificant role in the great plan.

Baobab

Energy flow food chain

The Web of Life

The natural environment forms a balanced relationship between all living things. No living thing exists in isolation, and in this way each is part of a community of living things. A tree standing alone on a hill is alone only in standing apart from other trees. In the soil below the tree, there are many other organisms which the tree relies upon for continued growth.

A forest is, therefore, a complex of atmosphere, climate, rocks, soil, water, plants and animals ranging from the microscopic ones to the large mammals. A complex of living things occupying a particular area is known as a 'biotic community'. This, regarded in combination with the interweaving non-living parts of the environment—soil, water, sunlight and air—forms an 'ecosystem'. Now if Man were to harvest a forest of trees commercially, he would consider not only the trees but also the parts of the ecosystem to which the trees belong. All living resources are inter-related. In any one place all these components function as a dynamically balanced whole called the ecosystem.

The term given to the study of the environment is 'ecology'. The word 'ecology' is derived from the Greek word 'oikos' which means 'house'. Ecology then is the study of the household affairs of plants and animals and how they co-exist in their natural state.

Water, soil, air, plants, animals and Man depend upon each other for existence and replenishment. Invisible threads or 'strands' bind living things to their environment. The complexity of these strands is illustrated in many ways. For example, soil is rendered fertile by animal droppings and rotted vegetation. Millions of tiny organisms in the water of seas, rivers or lakes provide food sources for fish and other water creatures. Shells, millions of years old, provide man with oil from which petroleum is made. The air we breathe is continually being renewed by the action of plants.

Each 'biome' or 'biotic community' supports a wide and varied cross-section of life. But each organism within a biome has its own particular pattern of existence; this place is called a 'niche'. Organisms adapt themselves to the climate of the niche, which contains the correct proportions of temperature, humidity, water, food and light. A scorpion, for example, is fairly sensitive to light and usually exists in a state of semi-darkness within a hollow log or under a rock. Here it preys upon insects which in turn require similar conditions. In this way, the log and the rock become the niche for the scorpion.

Each species has exclusive occupancy of its designated niche, although portions of the niche may house other types of animal life. If another species attempted to seek the same cover and food of another's niche, it would be competing directly and one of the species would become displaced.

A spider's web is constructed in such a way that the strands are intricately connected to weave a net, designed to entice the spider's prey.

Young salt-water fish exist close to the edges of lagoons or estuaries. Their presence provides food for countless fish, birds and reptiles.

A hawk feeds upon mice which rely on grass seeds for nourishment.

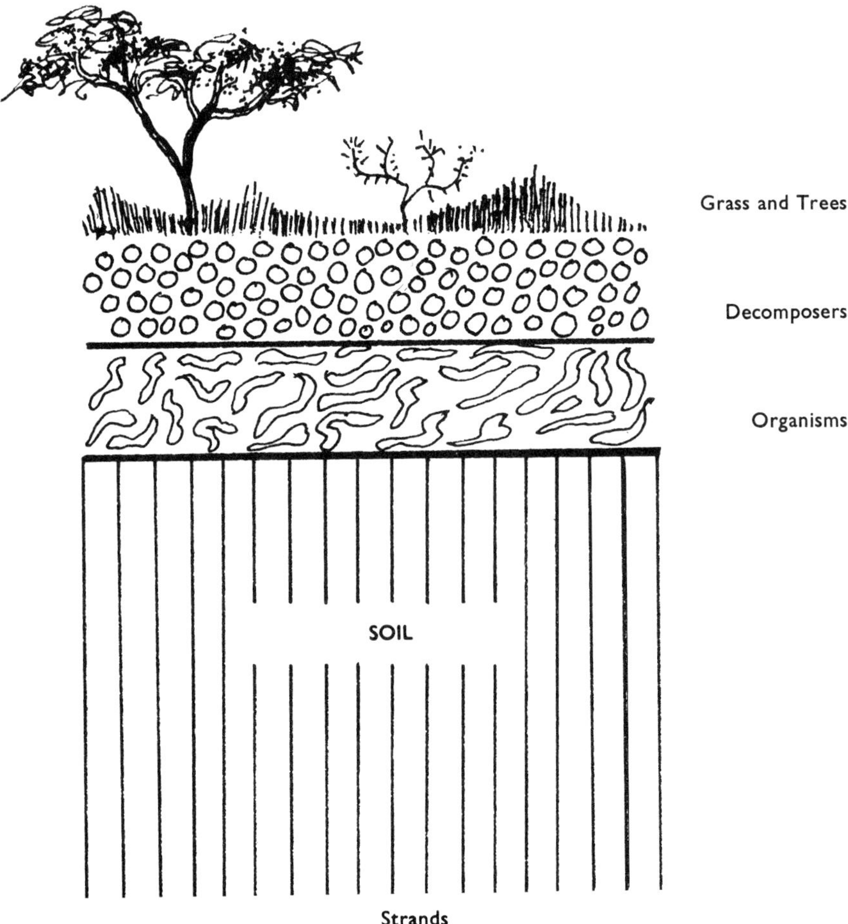

The hawk may support a number of parasites such as lice. On the death of the hawk, it will provide food for numerous organisms of decay. The mouse population will be eaten by such predators as owls and wild cats. The grass will be eaten by other herbivores. Food-chains, therefore, become interlaced into complicated food-webs.

In the food-chain and the flow of energy, plants are the only organisms which have the ability, through their leaves, of capturing energy from the sun 150 million kilometres away. With the aid of chlorophyll, water and minerals which are absorbed by the roots from the soil and transmitted to the leaves, carbohydrates and sugars enable the plant to grow. This process is known as photosynthesis.

A locust draws energy from the leaves of a plant. The locust is then eaten by an insect-eating bird which obtains its energy from the locust. A bird of prey captures the insect-eating bird and the energy is transferred from one living thing to another up and down the chain.

The amount of energy available in sunlight is large in relation to the amount actually captured and used by a biotic community. Through reflection and radiation, sunlight and energy often escape capture. In each energy transfer from plants to herbivores and carnivores, some energy is lost. Therefore, the quantity of green plant material produced needs to be large in relation to the quantity of herbivores that feed upon it. The energy stored in plant carbohydrates and proteins cannot be turned into an equal amount of animal tissue. A great amount is lost in the form of heat, chemical conversions that occur during digestion, and so on. Similarly, the quantity of herbivores exceeds the quantity of carnivores that depend on the plant-eaters for their energy supply. One kilogram of antelope meat cannot produce a kilogram of lion. Energy enters the system and is finally lost as heat is radiated. It is not cyclic as it has to be constantly captured to continue the system. The flow of all other life essentials is cyclic. It is never lost.

Patterns of different food-chains occur in the following:

Tiny water plants; tilapia (type of fish); barbel (fish); to crocodile.

From grass to zebra to lion.

(Hyena to jackals to vultures.)

(Flies—borers.)

Because Nature wastes nothing, energy is utilized to the very last resource. For instance, when the lion has finished with his kill, hyena, jackal and vulture take the rest, the hyena even going so far as to consume the bones.

Smaller things, like flies, feed off the decayed meat and lay their eggs there, so that young larvae can also derive necessary nourishment. Borers devour the remaining bones and the bones eventually disintegrate into the earth, enriching the ground in the process. All the animals, from the lowest forms of insect life to the large predators, enrich the soil in the form of manure. So the food-chain becomes a cycle as well.

Geology

The study of rocks, known to science as geology, is interpreted in volumes of books written on the subject throughout the centuries. Briefly, *Walk through the Wilderness* illustrates the Gondwanaland theory of how the southern African continent was formed.

It is widely believed that some 200 million years ago the southern continents were originally formed into one super-continent called Gondwanaland. This is known as the Continental Drift Theory. The super-continent gradually separated, forming the continents of Africa, South America, Australasia and Antarctica, which drifted to their present global positions.

Gondwanaland was surrounded by an unbroken chain of circumferential mountains which now occupy the individual continents as mountain ranges. The chain consisted of mountain ranges such as the Andes, East Australian and New Zealand. Great basins of sand and mud, hundreds of metres in depth, were found within the chain. The sedimentary slates and sandstones of the present-day southern continents were derived from the sand-basins of the Gondwanaland chain.

The southern part of South Africa was a large inland sea bordered by mountains during the Gondwanaland period. The process of erosion washed soil into the sea, forming swamplands which favoured plant and animal life. Here the prehistoric dinosaurs roamed. (The Greek word 'deino saurus' means 'terrible lizard'.) Remnants of fossils and prints have been found in Lesotho and the Natal Midlands.

With continued erosion, the inland sea became smaller, filling up with layers of sediment, scoured from the surrounds, known as molteno beds. Layers of sand, pebbles and boulders, known as red beds, formed our present-day sandstone. Plants and trees covered in this period were gradually turned into coal.

Eventually the inland sea dried up completely and the great reptiles disappeared because conditions became so unfavourable. For millions of years this basin remained a vast desert. Wind erosion created huge layers of sand. The same sand now forms the yellow and brown sandstone cliffs situated below the main escarpment in Natal known as the Little Berg.

Prior to the dissemination of Gondwanaland, the sediment layers were covered by hundreds of metres of basaltic lava. Each continent then assembled its load of sedimentary and volcanic rock after the dissemination of Gondwanaland some hundred million years ago. South Africa's portion of sedimentary rock is now called the Karoo system, which still underlies half of the country. Mountain ranges like the Drakensberg were created from the basaltic lava, interspersed with dolerite and inlaid with quartz and agate. The quartz and agate were formed during the last volcanic upheaval. Bubbles of gas were trapped when the volcanic materials solidified. Minerals, dissolved by water, percolated into the gaps and formed concentric layers of multi-coloured agate and quartz.

While the Gondwanaland continent continued to separate, sea water poured in to replace the disintegrated pieces. The land became more humid and intense rainfall set in, eroding and sculpturing the land to its present shape. South Africa is structured into different layers of older rocks and dehydrated deposits of the original inland sea. Then the sandstone layers of molteno beds occur followed by the softer red beds, windblown sandstone and a cap of basaltic lava some 1 500 metres thick.

The geological structures of South Africa have not changed during our lifetime, but from the above one can observe that the land differed markedly millions of years ago.

The geology of South Africa can be divided into three main categories of igneous, sedimentary and metamorphic rocks.

Igneous rocks are volcanic in origin, the Greek word 'igneous' meaning

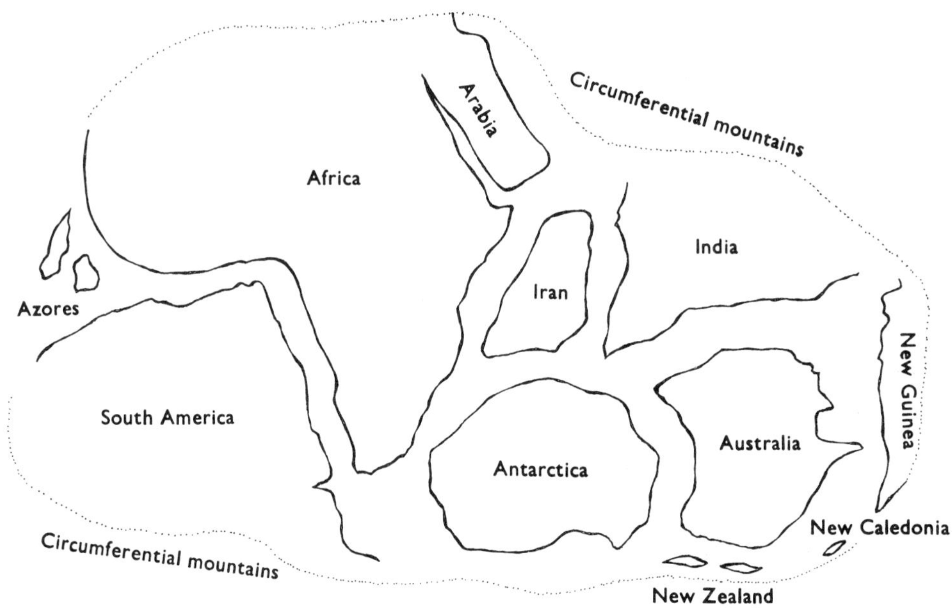

The super continent—Gondwanaland

fire. Basalt is dark green, uniform and fine-grained, frequently interspersed with pockets of agate and quartz. Basalt was once molten lava which poured into the earth's crust from volcanoes or fissures. Basalt is the most common igneous form. Associated with basalt is dolerite, which comes from a lava that never reached the earth's surface. Dolerite normally takes on the dual shape of a dike (dyke) and a sill. A dyke resembles a wall some 10 metres wide with a line of koppies at the surface as seen in the Karoo. Dykes appeared millions of years ago and were reshaped with constant weathering. A sill is a body of dolerite introduced along the layers of sedimentary rocks. In some areas a sill appears as a large mass up to 3 000 metres, for example, the Insizwa mountain of Griqualand East.

Granite is a coarse, granular rock comprised of quartz and feldspar. Granite varies in colour from grey to pink and is often used as ornamental stone. It sometimes glitters with specks of hornblende and black mica. Cape Town granite is intrusive while the granite rocks of the Limpopo and Zambezi river valleys developed from recrystallization, turning it into metamorphic or altered rock.

Sedimentary rock is derived from sediments deposited under certain climatic conditions. Tillite ('till'—a mixture of all grades and materials) comes from the action of glaciers, for example, Dwyka tillite. It is interesting to note that Dwyka tillite has been quarried in Durban for road metal and concrete aggregate for more than one hundred years. One can still study the quarry faces of the Lower Umgeni. The latter point underlines the Gondwanaland theory, substantiating glacier action where Durban now stands. Coal, originally formed from an accumulation of plant materials, now compressed, is another sedimentary rock of great importance.

Slate is a common sedimentary rock sometimes referred to as 'mudstone'. Minor escarpments in northern Natal are predominantly Ecca slate, which is used in the manufacture of bricks in Durban, Pietermaritzburg and Ladysmith.

The original characteristics of *metamorphic* rock were changed by deep burial in the earth's crust. Heat, stress and chemical reactions have now provided new characteristics. Examples of metamorphic transformations are: mudstone into slate, sandstone into quartzite, mudstone into lydiamite. Years ago, lydiamite was used in the making of stone implements. Some of the more ancient rock formations, altered metamorphically, are bound and known as 'gneiss'. Part of these rock formations are sedimentary in origin, the bonds caused by large proportions of silt build-up. Other formations are igneous in origin due to the process of lava flow.

The balance of the southern African rock formations comprises a combination of igneous and sedimentary, intruding along parallel layers.

Minerals are the building bricks of the earth, forming the components of most rocks. Minerals are defined as inorganic chemical elements or components found naturally in the earth. They include calcite, quartz, feldspar, mica, talc, and pyrite. Rocks vary in composition; for instance, two specimens of the same rock consisting of identical minerals but from different localities can have differing proportions of these minerals. Examples are granite, basalt, sandstone, chalk, slate and marble.

Earth's Natural Regions

With so many areas of the world differing markedly in appearance—the landscapes of rivers, seas and mountains—several natural environments, teeming with animals, are apparent. And each environment contains many forms of life adapted to a particular section of land, water, air, vegetation and other animals.

In animal life, vegetation forms a major factor for continued existence, and its features in any one area are determined primarily by climate. Ecologists divide the earth into ten main climatic zones, scientifically termed 'biomes', each supporting its own specially adapted plant and animal life.

To simplify the magnitude of the earth's special environments, the ten zones are classified beneath, so that the earth's major climatic regions can be perspectively understood:

POLAR REGIONS
CONIFEROUS FORESTS
TEMPERATE FORESTS
GRASSLANDS
DESERTS
TROPICAL FORESTS
MOUNTAINS
OCEANIC ISLANDS
INLAND WATERS
OCEANS

With deep contrast, vegetation differences are noticeable in polar regions, coniferous and temperate forests, grasslands, deserts and tropical forests. Mountains, where climate relates to altitude, harbour special animal and plant communities. Oceanic islands have distinctive animals due to their isolation, but have few climatic or vegetation characteristics in common. Inland waters and oceans comprise plant and animal life of aquatic system.

The zones are defined by patterns of rainfall and temperature and control their own seasonal changes. Combined with local soil conditions, these zones produce characteristic types of plant life, supporting characteristic animal communities. One combination of rainfall, temperature and soil, for example, has produced dense tropical forests in the humid equatorial lands. Another combination has created the arid deserts that cover about a fifth of the earth's land surface. The zone covering 70 per cent of the earth's surface, namely, the oceans, houses the greatest number and variety of animals, all of them ultimately dependent on the tiny plants of the plankton.

Because no species exists in isolation, no organism can be fully comprehended unless it is considered as part of its environment. The earth contains many habitats, each supporting its own highly adapted and integrated plant and animal communities. In any one area, all these components function as a balanced entity called an 'ecosystem'.

1. The spider's web upon which it depends for its survival forms the link in the spider's relationship with other living things

Photo: Dennis Cleaver

▲2. The mountain ranges of the Drakensberg were created from basaltic lava, interspersed with dolerite and inlaid with quartz and agate

◀3. Natural regions of the earth's environment contain many forms of life adapted to a particular section of land, water, air, vegetation and other animals

The Life of a Log

Decoratively speaking, a log performs a variety of domestic functions. Logs adorn drawing-room mantelpieces, crackle happily in open fireplaces, or create instant indoor gardens containing tiny tropical plants and variegated geranium leaves. Domestic exploitation of logs, however, is only a minor aspect of the part played by these seemingly inanimate objects in nature.

Throughout its life, a tree converts solar energy into wood fibres, withdrawing large quantities of chemical compounds from the soil. When the tree becomes a log, it begins to decompose, returning those compounds to the soil. During this period of decay, the log serves a number of useful natural purposes. There are sound reasons why an old tree log should be left lying in the veld. By its very size, the log conserves moisture in the soil beneath and it can also help to control erosion by halting the run-off of water.

Seeds of grasses, bushes and trees are frequently washed up or blown against the log, germinating in time under its protection. The plants which germinate under the protection of the log form the beginning of the herbivorous food-chain. Some of the seeds and fruits of these plants help sustain small birds and mammals.

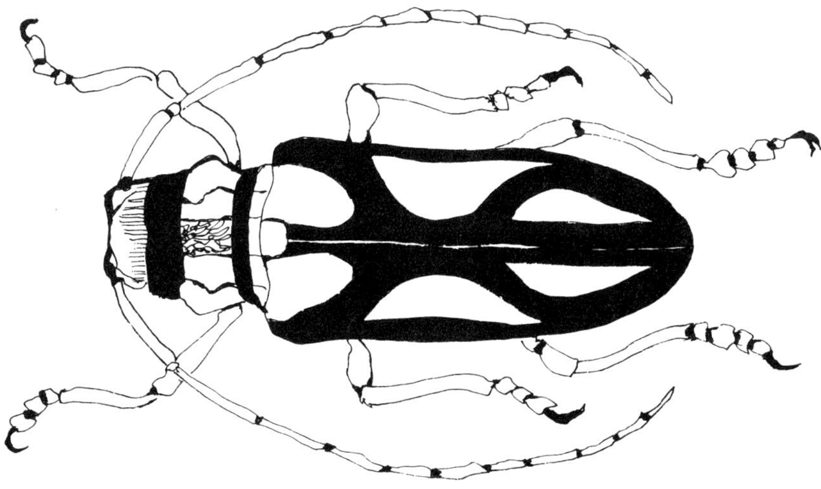

Long-horned beetle—*Cerambycidae*

Dead leaves and other plant material accumulate round the log, forming an absorbent sponge for water. The sponge eventually enters the process of decay, turning into humus and enriching the soil around the log.

Bacteria and fungi grow on the log and utilize remaining energy. The fungal threads form a network which penetrates the wood cells, decomposing the lignin and cellulose of which the log is composed.

The breakdown process continues with the log providing a niche for many insects. Termites, woodborers and beetles are common visitors to the shelter of the log; in turn, these and their respective larvae provide abundant food for bigger insects, scorpions, spiders, frogs, lizards—even small mammals such as the shrew, mongoose and aardwolf.

Snakes and birds prey upon frogs, lizards and smaller animals, which, in turn, fall prey to the various predators.

The log, therefore, provides an interesting and very necessary part of an ecological niche.

Nature's Laws

Neutralism

There are many interesting patterns that are followed by animals and plants living together in a balanced ecosystem. Some animal and plant species compete fiercely for food or living-space while others manage to work beneficially together.

Animal and plant species exist through various forms of association or interaction, for example, the indifferent state of 'neutralism' where neither species of plant or animal is ever affected by contact with the other. 'Competition' results in each specific population adversely affecting the other in the struggle for food, nutrients, living-space or other common need. In the antelope world, for example, nyala and bushbuck derive their nutrition from the same browse, but the nyala, being taller, can consume plants that the bushbuck cannot reach. In the plant world, many plants compete with one another for light, space, food and water.

Insect pollinating flowers

'Mutualism' is the system in which the growth and survival of both populations is benefited and neither can survive under natural conditions without the other. The terms 'co-operation' and 'symbiosis' are also employed to define this system. Many flowers rely on insects for pollination. In return, insects acquire nectar from the flowers as a main food source. A bee, for example, would never survive if he couldn't carry nectar and pollen back to the hive, the centre of the continuance of the bee populations. The bee then travels from flower to flower leaving behind some of the pollen, which enables the plant to reproduce. The bee and the flower are vital to each other for continued existence.

'Commensalism' relates to the benefit of one species while its counterpart, in terms of 'friendly association', remains unaffected. For instance, barnacles live in a permanent state of attachment to the hide of a whale but cause no harm or inconvenience to the whale. In the plant community, lichen is a combination of an alga and a fungus. The alga, which contains chlorophyll, produces food and energy; the fungus provides a 'skeleton' for the alga to grow on. Both the alga and the fungus are so closely associated that botanists refer to the lichen as an individual species. In nature, neither the alga nor the fungus could ever exist alone.

'Parasitism' occurs where one population has an adverse effect on the other through direct attack but at the same time is totally dependent on its host. Ticks, tsetse flies and mosquitoes are common parasites, while birds harbour many parasitic insects among their feathers. In the plant world, blight is a microscopic parasite found on a host plant, which it eventually destroys. A common parasitic plant, the loranthus, or mistletoe, is found clinging to the acacia trees which cover vast areas of the South African veld. These parasitic plants have reduced root systems modified into special absorbing organs for they do not need to absorb

Tsetse fly

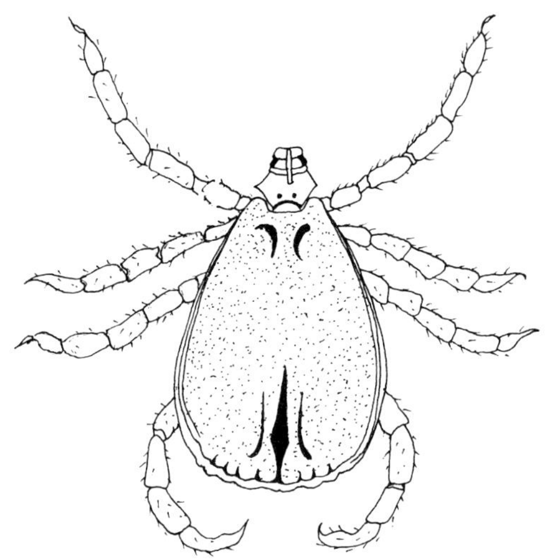

Tick—common parasite

'Strangler'

water or mineral salts from the soil. Their quantities of chlorophyll are smaller and either have scale-like and colourless leaves or no leaves at all for they rely on their host plant to capture the sun's energy.

If one ambles through a dense canopy-covered forest one is aware of many dangling plants which look like parasites because of the way they entwine themselves in and out of the trees. In point of fact these plants, known as 'epiphytes', are not parasitic but only use other plants as mechanical supports and do not draw their nourishment from these 'prop' plants. Like climbers, epiphytes abound in tropical and subtropical forests and fall into sun-loving and shade-loving groups, depending on the niche where they exist. Orchids, ferns and lichen are well-known epiphytes.

Common inhabitants of tropical and subtropical forests are the 'stranglers', especially the species of wild figs. At first, the seeds of strangling figs germinate like epiphytes, but their roots grow down, reaching the soil. As the fig grows, more roots develop, increasing in girth and surrounding the trunk of the 'host', which is gradually killed. The fig then becomes self-supporting, and while the host tree rots away, a hollow-trunked fig tree appears.

To understand the exciting process of how plants draw their nourishment from the elements, either independently or by relying on another plant, one must look at the process which controls the movement of energy converted into water and minerals from the air into root systems. 'Osmosis' is the term used to describe the movement of water containing dissolved elements through a membrane dividing a stronger solution from a weaker one, the movement being from the weaker until both reach the same level of concentration. This occurs in the roots of a plant. The soil water surrounding the particles of soil is actually a solution of mineral salts weaker than the solution in the cells of the root hairs so that water passes through the cell wall into the root hairs. From the root hairs, the water absorbed by them travels, also as a result of osmosis, into the wood-vessels of the root. Through the wood-vessels of the roots, stems and leaf stalks, it eventually reaches the leaves, then evaporates into the air. The process of osmosis affects certain plant adaptations. For instance, the plant *Scaevola thunbergii* or dune disc leaf not only stabilizes dunes on the beach but is also adapted to withstand salt spray from the waves—a very strong solution which would draw water out of the leaves if they were not protected by the very thickness of the leaf cuticle or waxy external layer of leaf.

The nivea plant is efficiently adapted to withstand salt spray because of a special silky coating on the uppermost surface of the leaves. If the

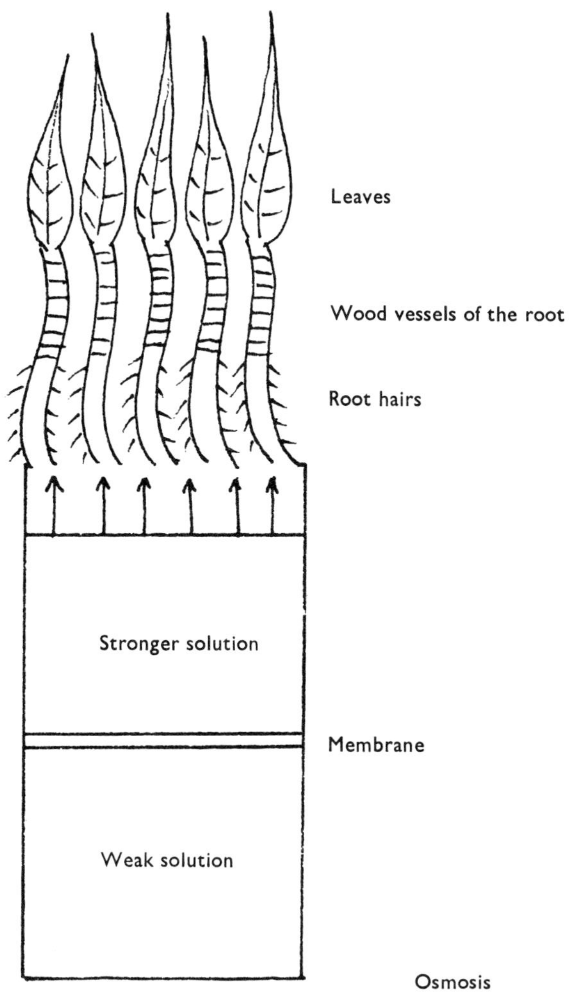

Osmosis

coating was not there, osmosis would occur in reverse, as the salty solution on the outside would be stronger than the solution inside the leaf and the plant would become dehydrated through excess loss of water.

Another interesting fact in the transportation of water and minerals from the air is apparent in the legume family. There are certain microscopic bacteria living in small nodules on the roots of plants of the bean family (Leguminosae), including the acacia species, which have the ability to extract nitrogen from the air of the soil. On their own, plants cannot achieve this. The nitrogen, converted into nitrogen compounds, reacts with the substances manufactured in the green parts of the plant to form proteins—important foods for animals and man. Agriculturists use leguminous crops to increase the nitrogen content in the soil.

The Water Cycle and the Ocean of Air

When water is warmed above a certain temperature, it turns to water-vapour, and can be moved by wind. The warmer the air, the more water-vapour is absorbed. When this air is cooled, some of the water-vapour condenses into fine drops of water (clouds) and under certain conditions comes down as rain.

The temperature of the earth and oceans rises through contact with the sun's rays. This results in a rise of temperature of the air coming into contact with the sun's rays, enabling that air to absorb more water-vapour and also to expand and become lighter. This lighter air rises whilst the cooler air sinks to replace it at the surface, resulting in a constant circulation of air. Naturally, the warm, moisture-laden air on passing through the cooler air condenses to form clouds, rain and snow, which grow heavy enough to fall back to earth, resulting also in a constant circulation of water.

As the earth revolves from west to east, so the air around the earth moves with it, but at different speeds, due to the curve of the earth from equator to poles. Warm air over the equator rises and then 'spins' off north and south towards the poles; conversely, cool air from the poles flows in to take its place. The result—a tremendous air movement, first towards the equator, then upward and in upper strata back to the poles.

Two factors produce the world's main winds—the spin of the earth and the movement of air between equator and poles caused by the heat of the sun. Ocean currents of different temperatures can also affect the air currents, as well as the changing seasons.

South Africa's main rain-bearing winds come from the south-east, depositing loads of moisture on the eastern seaboard and escarpment and leaving a relatively dry rain-shadow to the west of the country. The warm Mozambique Current has an important bearing on this. On the west coast, however, the cold Benguela Current, part of the Antarctic drift, washes the coast, resulting in cold air and very little water-vapour. Hence we have a moist east coast and a dry west coast. As a result, South Africa's eastern parts are well watered with many rivers having their sources on the main escarpment.

If one looks at a cross-section of South Africa, one can see that the land to the east of the escarpment falls sharply in a series of steps to the sea; this is especially so in Natal, South Africa's most-watered province;

whilst to the west the land falls gradually. This has a great bearing on the speed of run-off of the rivers flowing eastward into the Indian Ocean, resulting in deep valleys. This, coupled with the nature of the soil, also increases the occurrence of soil erosion.

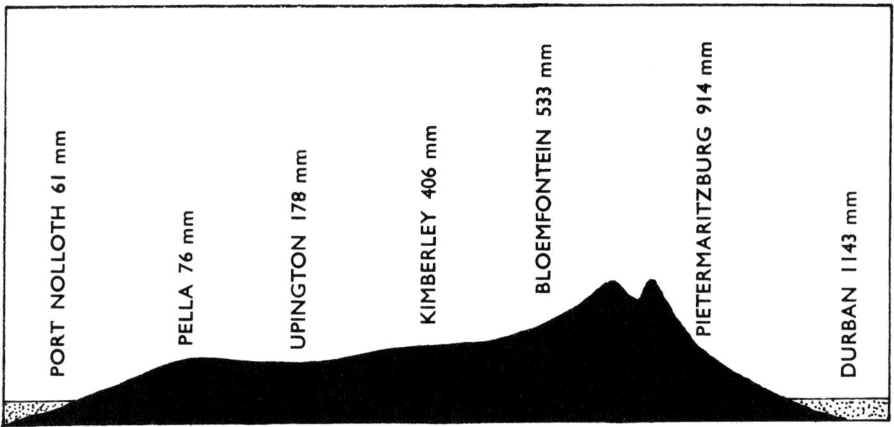

Diagram showing the decrease in rainfall from east to west in relation to vegetation

(1) Port Nolloth — coastal desert.
(2) Pella — The Karroos — desert shrub.
(3) Upington — The Karroos — shrub proper.
(4) Kimberley — Kalahari thorn veld.
(5) Bloemfontein — Highveld (grassland).
(6) Pietermaritzburg — Eastern Grasslands (bushveld to the north) Zulu.
(7) Durban — palm belt.

From Oceans to Mountains

COASTAL VEGETATION

Tides, wave action and currents are contributory factors in wearing away parts of the coastline. Ocean currents then deposit this sand elsewhere on the coast. There is always a taking and giving action. Hence, as mentioned earlier, plants play a major role in stabilizing the shifting sand on the beaches and forming dunes.

The first and most important pioneer of this shifting sand belt on certain parts of the South African coastline is *Scaevola thunbergii* (dune disc leaf). The Scaevola colonies form an open, scattered community up to about one metre tall. The Scaevola seeds are round and light, and roll down the preceding dune, where they are readily blown about . . . an interesting example of survival. The plant stem is capable of continuous elongation and production of adventitious roots from sand-covered stems. The plant thrives on being covered by sand, sending the energy to the growing-point. These communities halt moving sand, forming sand dunes, and as the plant grows, so does the dune. It has now established a launching-pad for its round spherical seed to roll down, starting off a new community. The plant is tolerant of salt spray due to a thick waxy leaf.

As we move inland from the sea, we find the Scaevola communities common, but now weak, as the dunes have been relatively stabilized.

Dunes stabilized by Scaevola

Other pioneer species now readily invade the dune where conditions for growth have been enhanced by litter and the improvement of soil conditions by the Scaevola. Once the dunes have been stabilized and there has been sufficient modification of soil by the pioneer species, shrubs species invade. As one walks further inland, so the shrub communities become denser and taller, terminating in canopy trees forming coastal dune forest.

As one moves inland, due to the protection of more sophisticated plant communities, an under-storey of grasses and smaller plants takes advantage of the protection and litter provided.

The older the dune forest, the more abundant the differing species of plants. The presence of the dune forest means that there is a great enrichment of the soil by humus.

Unfortunately, with man's invasion of the shoreline of most of our natural coastland areas, has come destruction of this type of succession and build up of forests.

Beach-buggies and landrovers have done much to stop the action of Scaevola taking place. Forest areas have been cut down for the production of crops. This leads in the long run to a depletion of soil content, ground cover and acceleration of erosion and wearing away of our coastline.

FORESTS

Nearer the tropics, where summers are hot and winters less cold, one finds warm temperate or subtropical broad-leaved forests of evergreen vegetation. These range from lush, many-storied forests in areas with high rainfall to drier forests and even scrub in areas where drought is common. The leaves in scrub forests have thick cuticles which minimize water loss during the scorching summer heat.

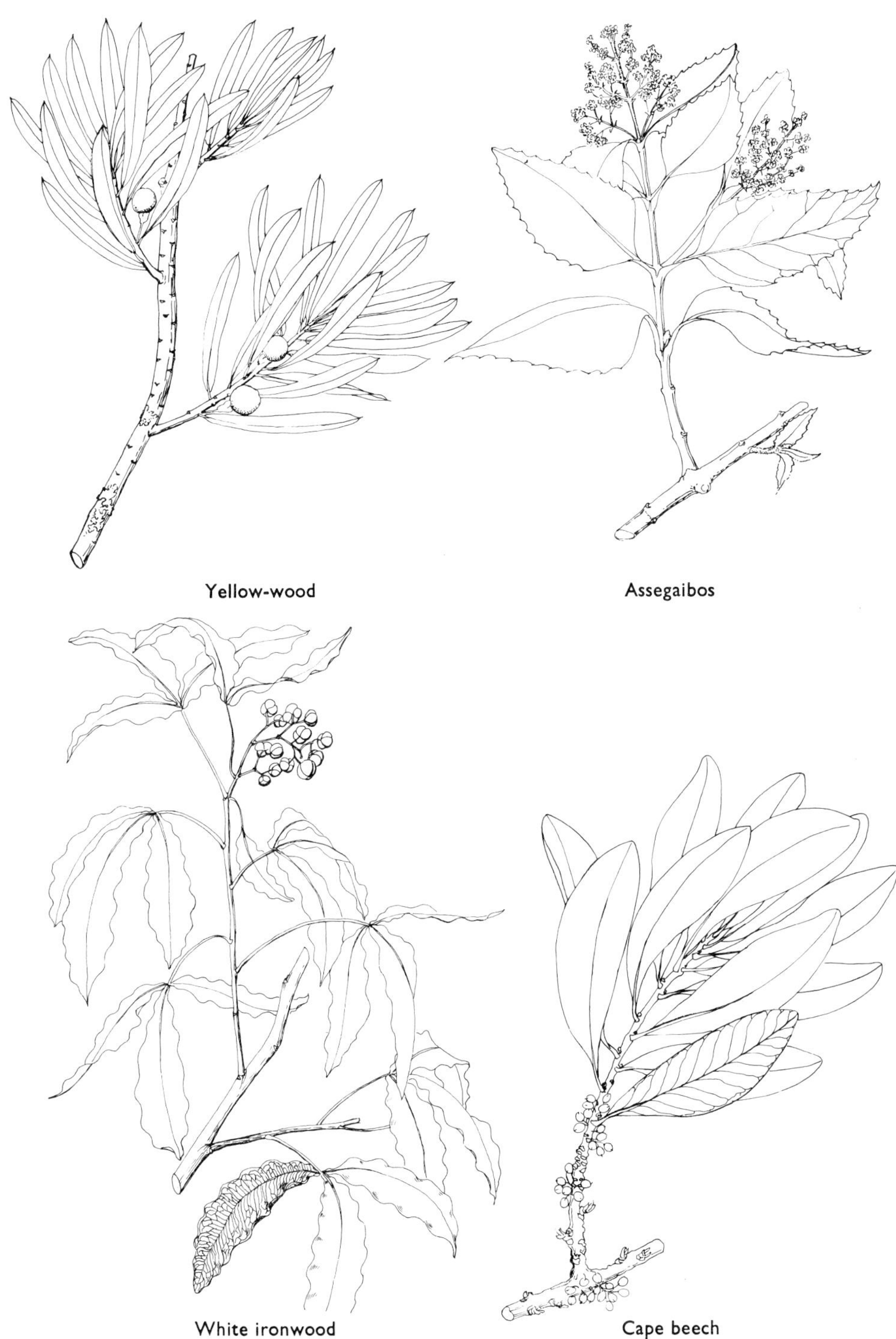

Yellow-wood

Assegaibos

White ironwood

Cape beech

Knysna touraco (*Touraco copythaix*)

In the Cape one finds the best example of a subtropical climate with dense shrubbery adapted to withstand summer drought. Trees bearing the protea are found up to 6 metres tall. In the eastern Cape the forests contain stinkwoods and Cape chestnuts. These forests grow on the lower slopes of the mountains and in the submontane shelf a few hundred metres above sea-level.

Bushbuck (*Tragelaphus scriptus*)

The finest remnants of this subtropical-type forest are at Knysna and Tsitsikama. They have survived because fires have been less severe and less frequent than in the drier mountain ranges. Inside, the forests are damp, cool and moss-covered with a base of leaf mould and ferns. The scroll-shaped leaves of many epiphytes trap water and form miniature pools which become the homes of water-insects and tadpoles. The forests are so thick that it would be difficult to see the legendary elephants of the Knysna forest.

The temperate forests of southern Africa, situated in the warmer latitudes, are predominantly evergreen, ranging from dense forests to parched scrub. Here a rich variety of plant and animal life abounds. The opportunity for the study of birds is limitless.

The largest area of temperate forests occupies the all-season rainfall zone from Mossel Bay in the west to the start of the Knysna forest in the east. Small patches of similar forest occupy some of the kloofs in the south-facing slopes of the Langeberg range to the west and the remnants of the cedar forests in the Cedarberg range.

In the eastern Cape Province interior, the Winterberg–Hogsback–Amatola escarpment is heavily forested on its southern slopes, and smaller mountains in the mist belts of the Transkei and Natal Midlands as far north as Nkandhla still bear remnants of forest growth. These forests include the evergreens of the tall yellow-wood (*Podocarpus latifolius*); the black ironwood (*Olea laurifolia*); the Cape beech (Myrsine—*Rapanea melanophloeos*); the assegaibos (*Curtisia faginea*); the stinkwood (*Ocotea bullata*); the sneezewood (*Pteroxylon obliquum*); the white ironwood (*Toddalea lanceolata*); and the kamassi (*Genioma camassi*).

In the forests, plant communities are formed with trees and plants competing for sunlight and shade. In temperate forests, pioneer trees offer shade for species requiring less sunlight and the montane forests are exposed to wind, which causes stunted trees and shrubs.

The forests, governed by conditions of climate and soil, adapt and form communities with many species of plants, mammals, birds, reptiles and insects.

GRASSLANDS

The extremes of temperature and wind in open conditions explain the great number of burrowing species of animals in the grasslands. Here, they are underground and protected against change in temperatures, storms and predators. The soil, even at moderate depths, is considerably cooler in summer and warmer in winter than the surface.

Termites

Termites build primarily according to weather conditions. The more exposed the region, the more the earth mound disappears and the bulk of the nest is in the soil. In open savannah, they burrow into the ground and cover their nests with domes or structures, using the excavated material with the help of their excretions. These hillocks are hard and solid and wind-resistant; they are also well insulated inside with air-chambers. In fact, some reptiles, such as some snakes, monitor lizards and tortoises, use these structures during hibernation, as they are so well ventilated and insulated. Snakes that burrow thus are adapted to dig, having modified snouts and shortened tails.

Rufus beaked snake

▲4. The log plays a vital role in the ecological niche

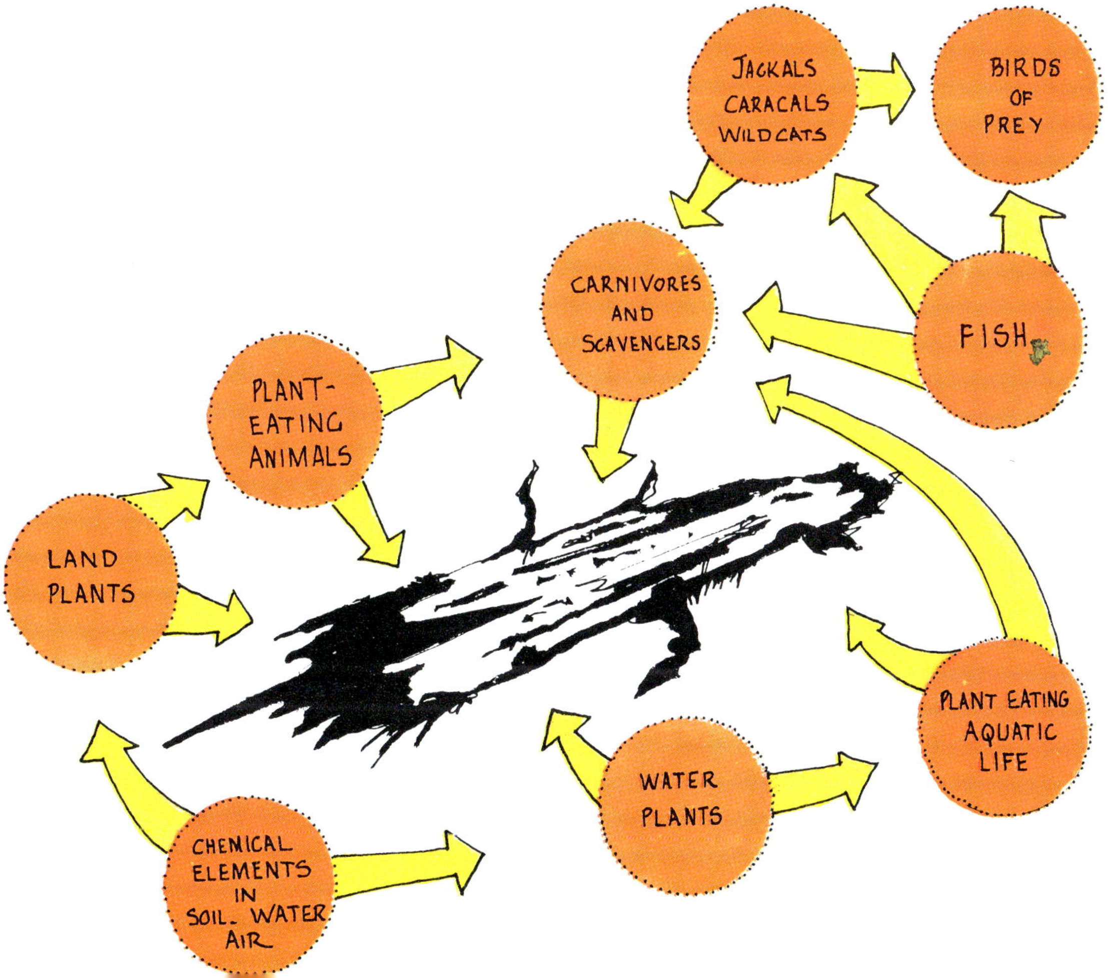

JACKALS CARACALS WILDCATS

BIRDS OF PREY

CARNIVORES AND SCAVENGERS

FISH

PLANT-EATING ANIMALS

LAND PLANTS

PLANT EATING AQUATIC LIFE

WATER PLANTS

CHEMICAL ELEMENTS IN SOIL. WATER AIR

▲ 6. The 'strangler' fig, using the host tree as a support, eventually killing it

▼ 7. 'Epiphytes' are not parasitic but only use other plants as mechanical supports

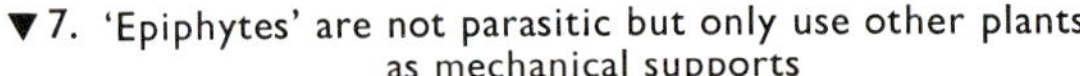

▲ 8. In the plant community, lichen is a combination of an alga and a fungus

Opposite

10. Temperate forest

11. Dune forest

12. Dune disc leaf (*Scaevola thunbergii*) the most important first pioneer of shifting sand-belts on certain parts of the southern African coastline

13. Plants play a major role in stabilizing shifting sand on beaches, thus forming dunes

▼ 9. Nivea plants are adapted to withstand salt spray with a silky coating

▲10 ▼11 ▼12

▼13

▲14. Plateau grassland

▲17. Birds adapted to survive in dry arid areas

▼15. Water. Vital to game reserves

▼18. Empty pan during the dry season

▼16. The bushveld, ranging from 150 to about 1 050 m above sea-level

▼19. After the first rains frogs appear

During dry seasons, when the last pools dry up and the rivers cease flowing, the larger animals migrate (if that is now possible due to fencing). Regular migrations of tropical birds occur during the dry seasons. A number of our African birds migrate north during our dry seasons, for breeding purposes.

With the first rains, a rapid change takes place. The grass shoots out and bushes and trees become green and begin to blossom. The countryside comes to life. The herd animals breed into flocks and families and are now widespread due to scattered water supplies.

The watering-places teem with life. The croaking of frogs is heard everywhere, tadpoles hatch out rapidly due to the warm water, and there is an abundance of food for the tadpoles—insect larvae and phylopods. The tadpoles' enemies are also present in great numbers.

Water-birds and water-beetles prey upon them as do terrapins and barbel, who have buried deep into mud during the dry periods. Frogs seem to appear from nowhere. In autumn, many species of frogs burrow into the ground or into holes in the ground to hibernate. Heavy rain after a long drought wets the soil and enables the frogs to dig their way out. In fact, it is easy to come to the conclusion that it has rained frogs in some instances as the ground can be literally crawling with them.

Tree-frogs' nests appear on the trees. These frothy-looking white bubbles adhere to the rock or tree overhanging the pool. The frothy mixture is excreted from the two parent frogs and then some 150 eggs are laid in the mixture. As the tadpoles develop, they wriggle and fall out of the nest into the water below to carry on their metamorphic cycle.

The winged sexual forms of termites and ants (flying ants) fly out of their nests to form new colonies. Insect-eating birds appear from nowhere to snap them up. Gnats, flies and grasshoppers appear. As the weather becomes more humid, the dung-beetles become active in the dung-heaps,

Dung-beetle

and one can see the feverish activity of dung-balls being rolled away by the beetles as a further food supply and nest for the new eggs. These balls are buried into the ground elsewhere and so serve the useful purpose of scattering and fertilizing the ground.

The grasslands are divided into different species, according to differing soil types and moisture availability. Much of South Africa's original savannah and grasslands, however, are much changed from what they were, due to the intervention of man.

Young nyala bull

Aristida congesta

The Ngongoni Veld

This occupies a narrow, irregular belt of rolling country just above the coastal forest belt. It lies on the slopes of the escarpment of the lowest in the series of plateaux of which South Africa is made up, and, lying as it does between 450 and 900 metres above sea-level, it is a great deal cooler and less humid than the coastal belt. The rainfall ranges from 750 to 1 250 mm per year. At lower levels, like the coastal belt, it is intersected by numerous bush-filled river valleys.

The sourveld has become dominated by the Ngongoni (*Aristida junciformis*) wire grass, usually a sign that this was veld or forest reduced by fire or man.

The surviving forests include the Nkandhla, Qudeni and Weza. Otherwise this veld is very open, except at the edges of the bush-filled valleys. It is regarded as a semi-intensive farming area.

Thornveld (mixed bushveld)

From about 150 to about 1 000 metres above sea-level—ranging from bushveld to Ngongoni veld.

Eragrostis spinosa

'Haak-en-Steek' (*Acacia tortilis*)

Montane Grassland

Alpine and subalpine fynbos and grassland form consecutive belts in this region. The alpine belt consists chiefly of dwarf Erica and Helichrysum (everlasting) and temperate grassland. The subalpine belt occurs on the escarpment slopes, from the basalt precipices down to the cave sandstone cliffs. The vegetation consists mainly of Erica–Phillippia fynbos, Themeda grassland and ouhout scrub.

Protea savannah occurs on the foothills.

Plateau Grassland (Highveld)

The climax over most of this area is *Themeda trianda* or rooigras. Overgrazing or faulty veld management results in a retrogression to species such as blousaadgras with a deeper root system, or further back in the succession to Aristida, including steekgras and kweek.

Fruit of the baobab

Themeda triandra

Rooigras is recognized as one of the best pasture grasses in South Africa and can be found in two forms—tall and short. The tall form grows most luxuriantly in the wetter areas with loose, deep soils; in the autumn months, the long leaves turn a reddish colour, hence the name. It is a typical sourveld grass.

Sourveld grasses in their young stages have relatively high percentages of nutritive elements in their foliage; at maturity, the proteins, carbohydrates and fats are mainly used for the formation of seed, and in autumn the minerals move down to the roots, leaving the foliage tough and fibrous. The increased formation of cellulose and true lignin in the autumn envelopes the nutritive elements left in the foliage. The red colour now indicates that the chlorophyll is being destroyed and the plant now becomes unpalatable to animals, and its high proportion of lignin and cellulose makes it rather indigestible—quite a contrast to its attributes in the first three months of growth.

Antelope of the highveld—Blesbok

The shorter rooigras grows generally in more fertile soils and in lower rainfall areas. It remains more or less palatable throughout the year and resembles a true sweet grass.

As we cross the escarpment and enter the drier western areas, the grass becomes less abundant, the firm tufts are more widely spaced and the leaves have a narrower appearance. Here, the types dominant are mostly Eragrostis and Sporobolus with Aristida becoming dominant in the broken Karoo veld, where the grasses intermingle with Karoo bushes.

This is a transition zone where the mixed grassveld is in constant defence against the ever-advancing desert activated through soil erosion and overgrazing.

Adaptation to open veld conditions

In general, only animals that tolerate dry air are best adapted to open country.

Reptiles and insects that can tolerate high temperatures and little moisture flourish here.

Birds survive in dry areas because their powers of flight bridge the gaps between water supplies and they also use water sparingly.

Mammals are represented by forms relatively independent of water due to the following body developments: rodents—cutaneous glands; antelopes—concentrated urine and dry faeces. For example, many rodents, such as porcupine, antelope, springbuck and ant-eaters, can survive for months without drinking water. The limited water taken in with their plant and animal food, together with water produced in the process of metabolism, is enough to satisfy their needs.

Gemsbuck

Elephants, except for herds with young calves, come to water only every two or three days. Zebra and black wildebeest will drink and thrive on alkaline water.

On the other hand, waterbuck, reedbuck, monkeys and even rabbits usually need water daily. Many animals of these grassland areas have adapted a keen sense of smell for moisture; in fact, distant rains can cause migration of herds of animals.

THE DESERT

The scorching heat of the day and the contrasting cold of the night make deserts the most extreme environments for animal life.

More than a fifth of the world's land surface, an area the size of Africa, is occupied by desert, semi-desert and arid regions. Sometimes a desert is a rocky plateau, sometimes a flat, yellow plain. Shallow, temporary lakes appear after rainfall, evaporate quickly, leaving a layer of glistening white salt. Other places comprise vast areas of shifting sand dunes, and elsewhere there are steep, rocky slopes. In the heart of the great desert regions, rainfall is scarce and the landscape has given way to the effects of wind erosion.

At dawn and dusk of every day, a desert comes alive. Mice, foxes, snakes and lizards emerge from their hide-outs to seek food. Most animals have, over the centuries, adapted to living with water-conservation techniques. A limited number of animals are able to survive in extreme desert conditions. Of these, many live in rock crevices and under loose stones. Running, jumping and burrowing vertebrates play an important role in the desert. Lizards are common among the running forms and mice are typical of the desert jumping animals. Rodents avoid high temperatures by remaining in their burrows until the coolness of the night. Ground temperatures during the day may reach 70°C or more; at night, if radiation to the sky is unimpeded, the surface temperature may fall well below that of the night air. The extreme temperature variations experienced at the surface diminish with depth.

Some species of desert rodents live without water and appear to exist entirely on seeds and dry plant material. Often this plant material is stored in the burrow, where, with increased humidity, the food stored takes up water and supplies the rodent with its water needs.

Antelope of the dry arid regions

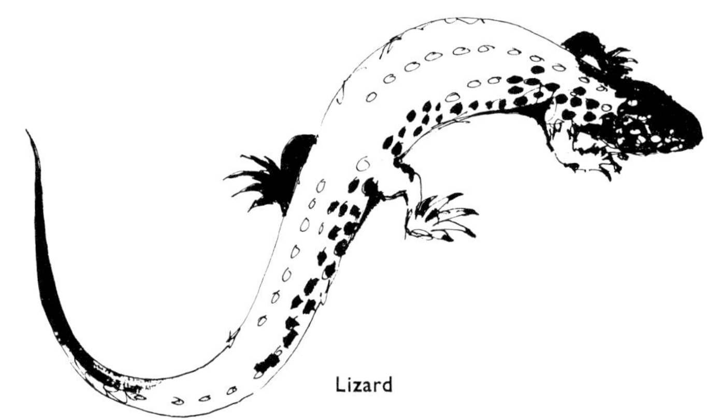

Lizard

Lizards and snakes are adapted for burrowing and are able to disappear quickly into the sand. The sidewinding adder (*Bitis caudalis*) produces horizontal movements with its body and disappears rapidly and silently into the sand. It has the ability to escape predators and heat. Reptiles devise other means of protection as they are incapable of ranging over great distances. Apart from burrowing away from heat, lizards have the ability to reflect excessive heat. Lizards can also change the reflection of visible light. It has been observed that when the body temperature of a lizard rises to an intolerable level, its skin turns lighter.

Colour is useful as a means of camouflage, too. Most desert animals tend to resemble the pale colours of their environment. In the desert, every bush and rock is an oasis of animal life.

Among the inhabitants of the sandy floors of the desert, some animals are adapted to prevent themselves sinking into the sand, enabling them to run or move more easily over the loose, fine sand. Some lizards, for example, have fringes on their toes. The sand grouse has feathery, tarsal toes; some rodents have well-developed hairs on the soles of their feet. The sidewinding habit of the snake is an adaptation to move more easily through soft sand.

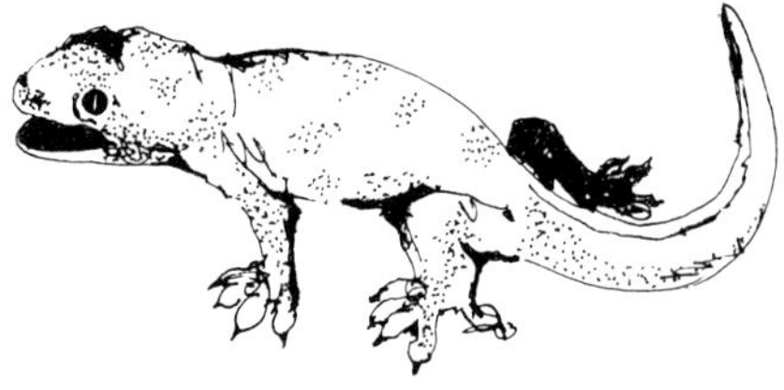

Gecko

The rainfall over the three Karoos is between 125 and 375 mm per year. The vegetation is remarkably uniform with slight modifications. This region is occupied by desert shrub of two types—succulent and woody, widespaced on bare, stony soil.

The Karoo succulent zone occupies the drier west of the Karoo area. The two most common bushes here are the 'kraalbos' (*Galenia africana*) and the geel melkbos (yellow milk-bush—*Euphorbia mauritanica*). *Aloe erox* is also found here. Tree growth is abundant along river courses, the most common trees being the *Acacia karroo* (sweet-thorn) and *Rhus viminalis*. Some trees are found on mountain slopes. These include the mountain plum (*Pappea capensis*), the gwarii (*Euclea undulata*) and the witgatboom (shepherd's tree) *Boscia albitrunca*.

Shepherd's tree

The Karoo bush proper or woody scrub covering the remaining Karoo consists mainly of gansies (*Pentzia virgata*), soetharpuis (sweet resin bush), *Euryops multifidus* and the saltbushes (Atriplex).

In addition to these perennial bushes, grass annuals spring up following rain. They soon die, however, in periods of severe drought. The Karoo bushes have a low rate of transpiration and are famous for their resistance to drought.

One often wonders how sheep manage to survive so well in desert conditions, but due to the low transpiration rate and deep root systems of the bushes, they are able to outlast any of the grasses in drought conditions. These shrubs maintain a high mineral content, rich in protein,

even during the dry season. They are specially rich in phosphates, potassium, sodium and chlorine.

THE DESERT SHRUB

To the north of the area occupied by the Karoo bush proper lies the desert shrub type of vegetation. This extends to the border of the Kalahari and across the Orange River into South West Africa.

Here, in the sandy, granite plains, grows the driedoring (*Rhigozum trichotomum*), reaching a height of one metre. We also find the pomegranate (*Rhigozum obovatum*) and the Bushman grass (*Aristida brevifolia*). *Aloe dichotoma*, the kokerboom, is common on the slopes.

Kokerboom

From the mouth of the Olifants River in the western Cape stretching north to Angola, the semi-desert conditions in the south intensify to true desert in the Namib of South West Africa, with a rainfall of 50 mm a year.

The southern portion of this belt is often covered in mist, which encourages a good cover of low, succulent bushes. The succulent zone extends inland to the lower Orange River valley.

Northwards of the Orange River in the coastal desert, the succulent bushes gradually thin out and are replaced by a severe desert surface of sand dunes, where there is little or no plant growth at all except for a few isolated patches of steekriet (prickle reed, *Eragrostis cyperoides*) or ostrich grass (*Eragrostis spinosa*), and welwitchia.

Eragrostis cyperoides

Desert landscape

Survival Techniques of the Bushman

In terms of conservation the Bushman was an early pioneer, for then both man and beast were equal. Like most creatures of the wilderness, the Bushman killed to eat, and although the game virtually belonged to him, he never destroyed the herds, for without them his own survival was limited. With his famous rock-art, the Bushman recorded for posterity the story of his own life style, his rituals, customs, culture and mythology, and also the animal herds that once roamed the land.

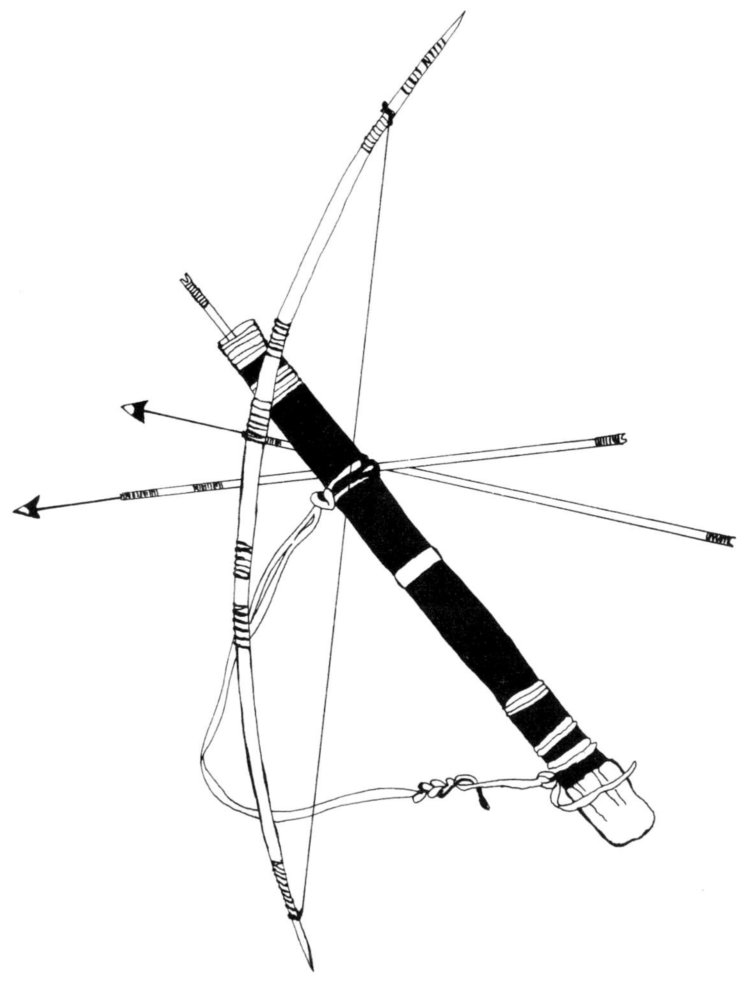

The presence of the Bushman in southern Africa dates back twelve or fifteen thousand years. Bushman life has now disappeared from the Republic of South Africa, Rhodesia and Zambia, and is today confined to South West Africa, Angola and Botswana.

Bushmen of the Kalahari

The greatest numbers are found in Botswana, where they live in the more remote areas, existing on a subsistence level and living side by side with nature.

Acacia Karoo (sweet thorn). Common tree of the dry Karoo

The Namib. Coastal desert of South West Africa

▲22. Deserts cover more than a fifth of the world's land surface

▲23. The Brandberg, S.W.A.

▼24. Kalahari Desert. Home of many species of wildlife

▲25. Part of the diet of the Bushman—the steenbok

◀26. The presence of the Bushman in South Africa d
back 12 or 15 thousand years

Opposite

27. Cheetah on a springbok kill in the Kalahari Gems
National Park

28. Cheetahs attain high speed when running down t
prey

27

Photo: Alf Pexton ▼28

Photo : E. Shelwell

▲29. Baby white rhinos awaiting translocation

◀30. Many insects have bright colours to merge into th
respective backgrounds

Overleaf
33. The African bushveld
34. The leopard—an elusive inhabitant of the bushv
35. Many species of Acacia
36. Impala lily

Opposite
37. Animals of the bushveld

▲31. Vegetarian mammals (herbivores)

▼32. On trail. The field officer explains the interacting cycle of a 'midden'

▲33 Photo: Alf Pexton ▼34 ▼35

▶36

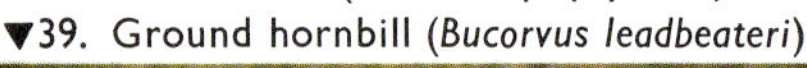

▲38. Waterbuck (*Kobus ellipsiprymnus*)

▼39. Ground hornbill (*Bucorvus leadbeateri*)

▼40. Giraffe (*Giraffa camelopardalis*)

There is sufficient historical evidence to assume that the Bushman, whatever the present situation, was originally a hunter–food gatherer. Although the Bushman living in the more arid regions of Botswana has suffered some cultural restrictions imposed by the environment, the hunter–food gatherer has developed an intricate association with his environment, which has been ecologically unaffected by his occupation over many years.

The widely scattered distribution in southern and south-central Africa of rock-engravings and paintings executed by the Bushman shows that this race once covered most of the southern half of the African continent. The records of early travellers, dating back to the fifteenth century, indicate the existence of the Bushman along the coast.

The Bushman's place in the ecology of the reserves is that of predator on the ungulates, springhares, jackals, foxes, rodents, birds and insects which they hunt and eat. Bushmen are also the rivals of browsers and fruit-eaters for the edible plant foods.

Bushman art

In collecting plants with inedible seeds, the Bushmen act as seed distributors, although the importance in this role is less than that achieved by mammals and birds in the act of seed distribution. In their utilization of bulbs and tubers, they are careful not to strip an area completely but always leave some plants intact as seed-bearers for the next season.

The Bushmen are never wasteful in their hunting and kill only as much as they require. Their hunting methods, which include bow and poisoned arrow and the use of snares, involve a minimum disturbance of the herd to which the hunted animal belongs. In fact, the disturbance caused by a Bushman hunter is far less than the furore created by a lion or leopard making a kill.

The diet of the Bushmen is predominantly vegetable, and even at the height of their hunting activity meat accounts for less than 40 per cent of their food intake. They are principally dependent on much the same food as the game animals and their energy and activity through the seasons are closely related to that of the animals.

So in tune with nature are the Bushmen that in times of drought, when the animals are weak, the Bushmen are also listless and unable to take advantage of the animals' debility to hunt them more easily. When the Bushmen are well nourished and have good reserves of energy for frequent hunting, the herds themselves are also in good condition and have a wide choice of grazing which enables them to move to other tracts of country. The Bushmen, however, are not free to follow game for unlimited distances, as they are confined to the territory of the bands. As hunting outside of their territories is disallowed, they are careful in the handling of the herds.

In an average year, the number of animals killed by one tribe is

Diet of the Bushmen

approximately 1 200, which includes kudu, eland, gemsbok, hartebeest, wildebeest, springbuck, duiker, steenbuck, springhare, porcupine, warthog, fox, jackal, rodents, birds, tortoises, snakes, ants and termites.

Out of that 1 200, they hunt over 400 tortoises, 300 rodents and 300 springhares. Bushman hunting has not been sufficiently intense to cause any disturbance of the ecological balance and they are, therefore, an integral part of a fairly stable ecology.

Bushmen have lived and hunted in their present areas for over a century, and in that time have not reduced the game herds. As long as they continue their present methods of hunting and food-gathering and are not subject to outside influences, it is unlikely that they will ever have any adverse effect on the animal or vegetable life of the reserve areas they now inhabit. In fact, they perform the highly useful function of culling the game.

Though primitive, the Bushmen utilize the assets of their environment to the mutual benefit of themselves and the areas from which they derive their livelihood.

Bushmen rely on the unimproved, natural resources of an area, making no attempt to improve and stabilize its productivity by domesticating animals for milk, meat, skins or wool, nor do they stimulate the growth of the edible plants by cultivating them. They are at the mercy of the elements and the resultant effects on their habitat. Good rains create a profusion of plant growth to satisfy both man and game, which, in turn, means availability of meat. Drought destroys plants, and the subsequent misery of hunger and thirst is often followed by the ravages of disease which affects Bushmen and animals alike.

Food-gathering, which is chiefly the work of women, is a more important subsistence activity than hunting, for, as we have seen, the Bushmen rely more on vegetable foods than upon meat. Among the Bushmen, there is no competition for food plants. Even when a special plant is scarce, competition is suppressed by the pervading insistence on co-operation, and women will ensure that enough plants exist for food-gatherers from another tribe. The daily pattern of work is dictated by the season and depends on the availability of various species and on the weather.

Bushmen display much imagination in the utilization of food plants, according especial attention to those which yield the highest nutritional value and give the greatest degree of satisfaction. For example, they collect various types of roots, seeds, berries and leaves which are richest in fat, protein, starch, vitamins and sugar. Berries are crushed, roots are roasted and wood fibres ground and chewed for maximum nutrition and appetite appeal.

The Tree of Life

The old adage pointing out that 'beauty is in the eye of the beholder' was never truer than in its application to thoughts on a wilderness experience.

Such is the vast variety of wildlife—be it flesh, fish or fowl—that it is surprising for so many people to have particular favourites. The exciting panorama of alert game at a waterhole, for instance, may well be lost on an ornithologist focusing all his attention on a fish eagle or a lilac-breasted roller; and the skin-diver in his subterranean world of wonder wouldn't thank you for a ringside view of a cheetah kill.

In this pursuit of the specific, however, it behoves one to have a basic awareness of the general order of the Animal Kingdom—if only to recognize the incredibly broad spectrum that's there for the seeing and hearing.

Beauty is in the eye of the beholder

The Classification of Animals

Various common names are often given to the same animal in different areas, hence the only reliable name that can be accepted by scientists from all over the world is the animal's specific or 'Latin' name. This consists of two names, the genus or generic name and the species or trivial name. This of course goes a bit further. Many species are grouped into genera, genera into families, families into orders, orders into classes, classes into phyla and all phyla of animals are grouped into the Kingdom Animalia.

For example, the species is the basic unit of classification of organisms; it is a kind of living organism, e.g. a domestic cat (*Felis domesticus*).

A genus (pl. genera) is a collection of closely related species, e.g. domestic cat (*Felis domesticus*), the wild cat (*Felis lybica*) and the serval cat (*Felis serval*). All belong to the Genus Felis.

A family is a group of closely related genera, e.g. the cheetah belongs to the genus Acinonyx, the leopard to the genus Panthera, and the domestic

Greater kudu: *Koedoe*
Tragelaphus strepsiceros Pallas

cat to the genus Felis. All, however, belong to the cat family or Felidae.

An order is a group of closely related families which have certain traits in common. For example, the cat family (Felidae), the dog family (Canidae) and the bear family (Ursidae) are all flesh-eaters, so all three families are grouped in the order Carnivora.

A class is a group of related orders. All of the aforementioned animals are hairy animals which suckle their young, as do gnawing forms such as rats, guinea pigs (order Rodentia) and human beings, apes and monkeys (order Primates). Thus these related orders are grouped in the class Mammalia in contrast to the classes Reptilia, Amphibia and Aves (birds).

A phylum is a group of related classes, e.g. all the animals mentioned have backbones and are included in the phylum Chordata. This shows that they are different from the phylum Mollusca, which includes snails, mussels, or the phylum Arthropoda, which includes insects, crabs, spiders and scorpions. All the animals belong to the Kingdom Animalia (Animal Kingdom).

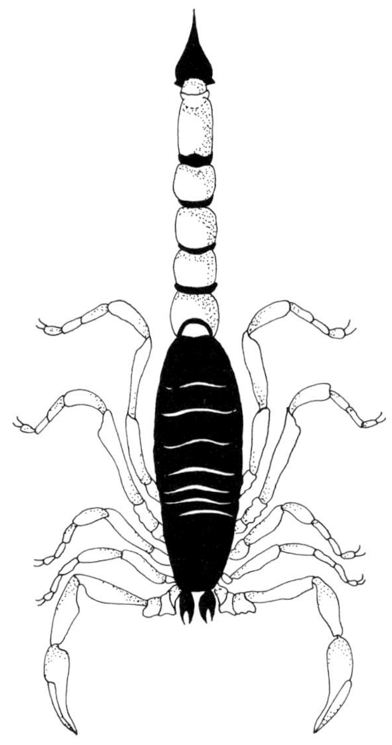

Phylum arthropoda. Class arachnida

Animal Adaptation

The living organisms of this world with the greatest survival rate are the ones which have adapted themselves most effectively to their surroundings.

In the animal kingdom there are many ways in which animals have developed skills, resulting in highly specialized forms of adaptation.

When one is lucky enough to spot a Kruger Park lion, one is truly amazed at the inconspicuousness of the beast—a tawny-looking creature in a maze of tawny-coloured grass. While, on impulse, it appears so clever, the ability of the lion to camouflage itself, by blending with the colours of the environment, is a means of natural protection. Perhaps the most dramatic example of animal camouflage is the chameleon, which changes the colour of its skin continuously as it moves warily about the environment.

Chameleon

The antelopes of the bushveld have adapted their limbs to move at great speed so that when danger threatens, either in the form of man or predator, they can escape. Predators such as the cheetah have adapted their bodies to run at phenomenal speed when hunting in the open savannah.

Animals such as antbears have developed long, sharp claws to dig into termite nests and to dig out burrows in which to live. The burrows are frequently taken over by warthogs, hyenas and porcupines, who clean the dens and use them for shelter and protection, although they feed above ground.

In the breeding season, these dens afford ample shelter for the animals' litters. Animals such as hares seek food above ground but dig their own burrows.

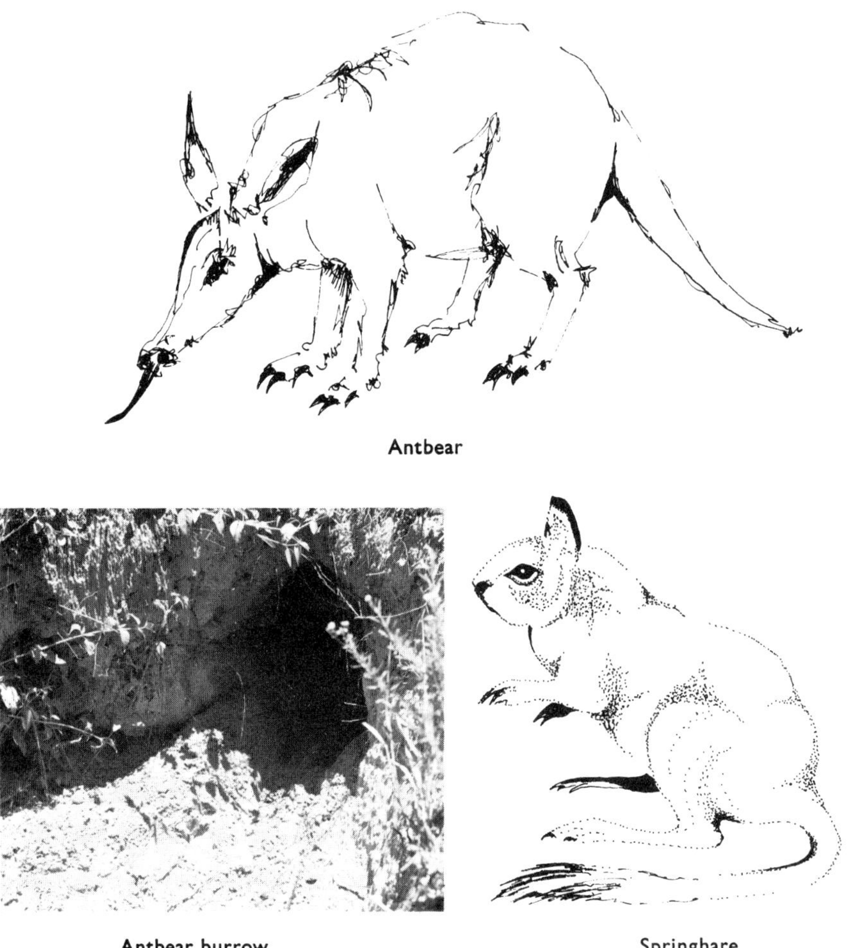

Antbear

Antbear burrow

Springhare

Some animals with exceptional eyesight, living in the tall grasslands, have the ability to jump very high, and are thus able to see their predators at a considerable distance. The little springhare of Africa protects itself in this way. The kangaroo and wallaby of Australia jump for different reasons; their front legs are adapted to pluck and to feed and they move in an upright position, developing strong jumping rear legs.

Although flying is confined to birds and insects, there are a few tree-climbing animals and lizards which have adapted flaps from neck to shoulder, enabling them to glide from tree-top to tree or ground. Examples are the flying squirrel and a lizard called the flying dragon.

The bat is the only mammal which has adapted itself to life in the air. Many animals have the ability to seek food or shelter from predators in the trees.

Aside from fish, animals such as dolphins, porpoises and seals have successfully adapted themselves to life in the water—their bodies resemble those of fish with flipper limbs and no external ear. Some birds, such as cormorants, dive and swim under water, while turtles have adapted to life in the sea. Crocodiles use their tails as the primary means of propulsion in the water. Most mammals are able to swim.

From the simple five-fingered reptilian hand, the fore limbs of mammals have developed according to their different ways of life. For example, the fore limbs of the ape have been developed for grasping:

Man—for holding and manipulating

Flesh-eater—catching and tearing	Camel—walking on soft ground
Antbear—digging	Horse—running on hard ground
Seal—swimming	Bat—flying

Birds, too, have adapted their feet according to their way of life, for example:

Swimming foot—in the case of ducks, geese, gulls, pelicans
Foot for walking and scratching—fowls, partridge
Perching foot—crows, larks and most small birds
Climbing foot—woodpeckers, parrots, cuckoos, mousebirds
The beaks of birds are also specially adapted as in the case of:
Finch—thick bill for cracking hard seeds
Swift—a wide gape for catching insects in mid-air

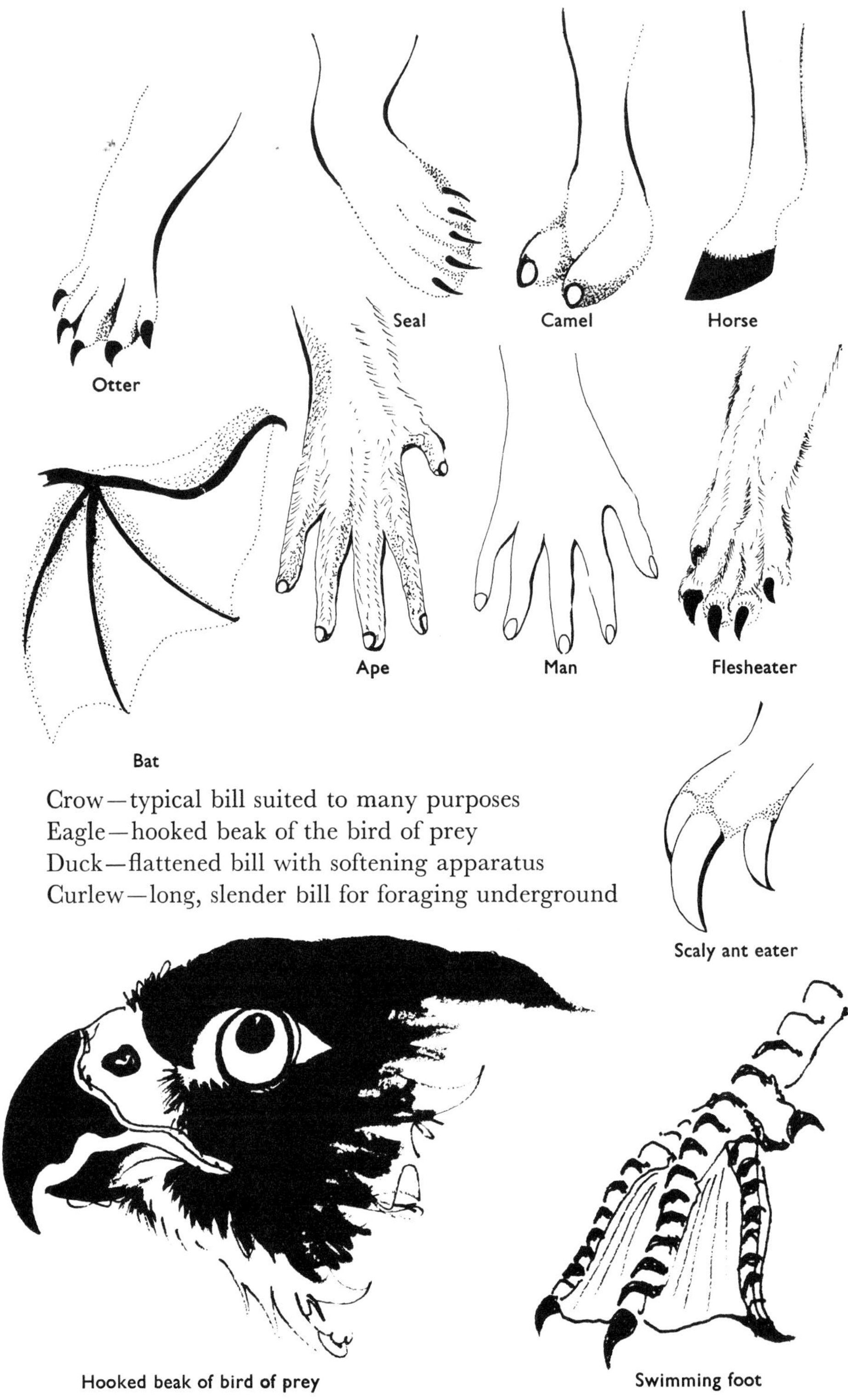

Crow—typical bill suited to many purposes
Eagle—hooked beak of the bird of prey
Duck—flattened bill with softening apparatus
Curlew—long, slender bill for foraging underground

All biological systems are designed to maintain the so-called balance of nature. This state of perfection is never actually achieved as there are too many environmental factors continuously influencing the system.

The natural process continues with biological factors which control population growth, for example, disease, predation, accidents, climate, food supply and stress.

Where these factors are not fully operative because of man's influence, for example game reserves, man has had to give nature a helping hand,

Elephant culling and white rhino translocation

namely culling game and translocating game by means of scientific game capture methods. It is very easy to imagine how the total environment would suffer if there was no biological control of the population size of locusts, for instance.

The diversity (variety of species) within a community reflects in part the diversity of the physical environment. The greater the environmental variation, the more numerous the different species. This is because there are more microhabitats available and more niches to fill. To illustrate this phenomenon would be easy if one compares the number of different

species occurring in the Canadian tundra as opposed to those found in a tropical rain forest.

In terms of conservation, it is vital that we conserve the species' diversity as a whole ecosystem and not select random aspects of the ecosystem in a particular region. A classifical example of the past was the predator control conducted in the game reserves of Zululand. In the mid-1930s, when the mistake was recognized, this practice was stopped, but the following predators nevertheless became extinct in Zululand—the lion, cheetah, brown hyena, African wild cat and the hunting dog.

African hunting dogs

Mammals, as their Latin name tells us, comprise a class of animals the females of which produce milk to feed their young.

Probably the class with the most general appeal, this immensely varied group of species is endowed with outstanding adjuncts to survival—in many ways superior to other vertebrates.

Being 'warmblooded', mammals have hides or skins, fur or hair as well as fat, which insulate their circulations against changing climate, temperature and local environment. When hot, they can sweat; when cold, they can shiver. They can cool themselves by panting (like canines), or just maintain sufficient body temperature for bare existence when hibernating. Their adaptation is extremely well advanced.

Large ears for hearing

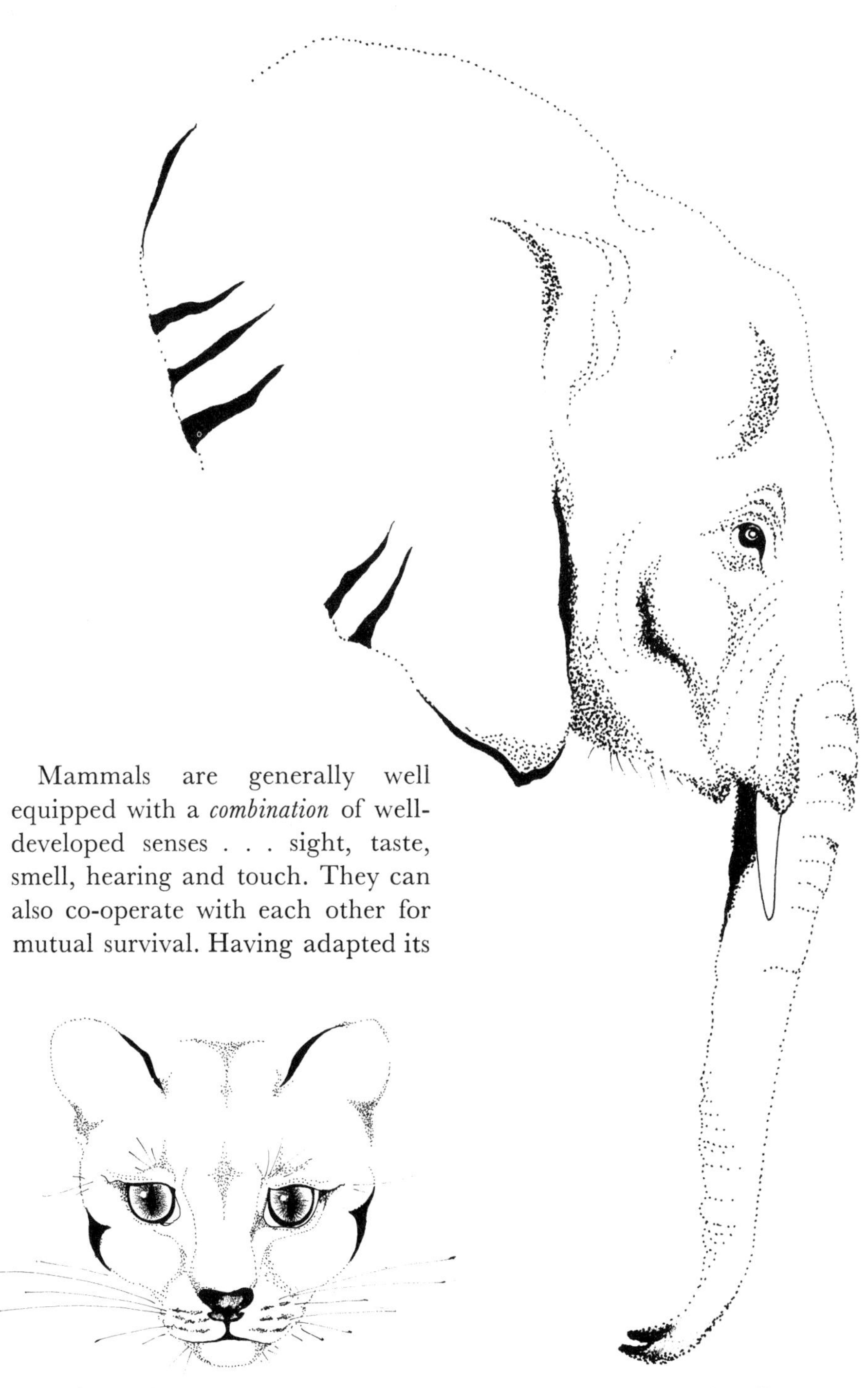

Mammals are generally well equipped with a *combination* of well-developed senses . . . sight, taste, smell, hearing and touch. They can also co-operate with each other for mutual survival. Having adapted its

Sight and touch spots

Hearing and smelling

species to underground conditions, the mole is virtually blind, but its highly sensitive nose guides it to food and warns it of danger. Mammals living in arid or actual desert conditions can go for months without water. And life in the thick bush and long grasses of the savannah calls for superb hearing . . . as witness the king-size ears of the kudu and steenbuck, for example.

Though by no means exclusive to the class, camouflage is also a feature of the mammal's survival pattern. It is a never-ending source of wonderment to the wilderness visitor that creatures of the magnitude of elephant and giraffe can merge so expertly into the locale. Only when they move is the camouflage rendered almost useless.

Animal coloration

Most mammals live in a dull, grey world. Only a few—the apes, for example—have colour vision. Others—usually animals that are active at night, like the cats—are famous because their eyes glow in the dark. The eyes, however, do not generate any light themselves; they merely catch up and reflect whatever light is present in the gloom. Behind this remarkable power lies an interesting fact. The inner wall of the eye is coated with a substance called 'guanin'. This has a metallic lustre of silver or gold and brightens dimly-lit images on the retina of the animal's eye so that it can see them more sharply.

Nerve fibres end in tiny raised points on the skin, and are known as touch spots. A touch spot usually has hair on it. The hairs themselves are not sensitive but act as levers to press the touch spot. Long hairs or whiskers on animals are connected with touch spots.

Some animals have bright colouring to merge into their respective backgrounds, but animals of the open grasslands are noted for their lack of markings. The underparts of these animals are usually white, or lighter than the colour of the back. The brightest light comes from above and lightens the back, but throws the underparts in shadow, so that both upper and lower parts look alike; the outline of the animal is thus obscured in certain lights if the animal is motionless.

African elephant—largest land mammal

41. African cheetah (*Acinonyx jubatus*) racing to extinction

▲42. Wave action Photo: A. Gifford  ▼43. Seen on east coast zone

▲44. Lake St. Lucia Photo: Paul Dutton  ▼45. Hippo are important to the ecology

How Mammals Live

IDENTIFICATION OF MAMMALS BY SKULLS

It should be possible to identify mammals by their skulls when found. Teeth provide valuable clues as do size and position of eyes and length of nose.

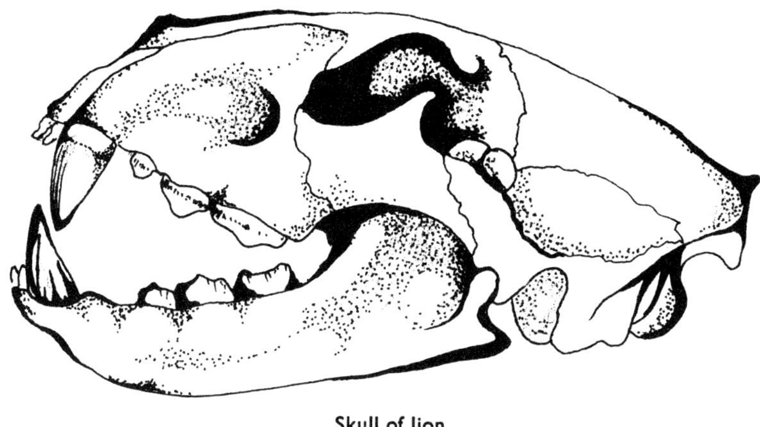

Skull of lion

For instance, an animal with a long nose will have a keener sense of smell than those with a short one. Eyes placed in the front of the head are of animals that have good eyesight and are usually predators. Those with eyes on the side of the skull are usually prey species as they depend on this type of eyesight for attack in the rear.

From teeth we can see whether an animal is a carnivore, herbivore or insectivore, whether it is a browser or grazer.

FEEDING

Vegetarian mammals (herbivores) are usually of a large size and therefore need a large supply of food to keep them going. To achieve this large intake, the animal's life pattern is transformed into almost one long meal.

The problem of obtaining food is usually more complicated for the carnivores. Γhey have to work for their food by hunting, and greater demands are made on their intelligence to survive.

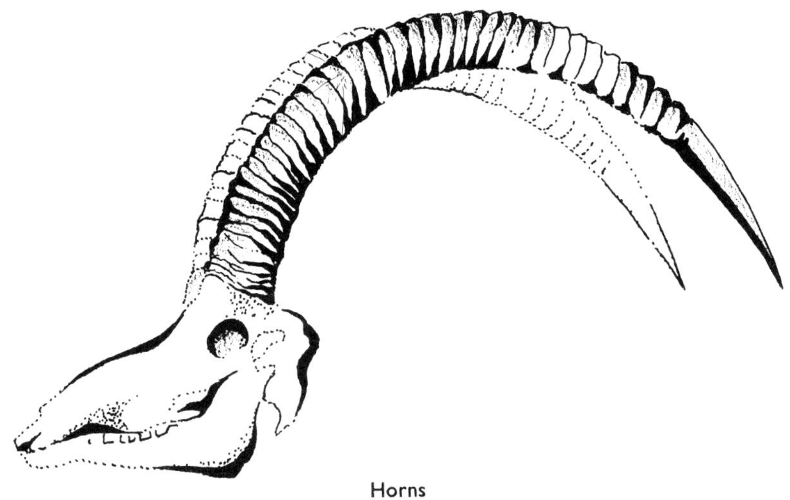
Horns

ATTACK, DEFENCE AND SURVIVAL

The battle for survival makes it necessary for all mammals to have offensive and defensive adaptations. Examples are given in illustrations, e.g. horns, armour-plating, quills, colour variation and change in coats.

Armour

Porcupine attack

TERRITORY

Territorial behaviour of animals has become a fairly recent and interesting study amongst ecologists of late. Where before it was believed

Territorial impala

that food was the salient factor controlling the dominance within a species, it has been proved fairly conclusively that territory is the important factor.

It would appear that all living things revolve around this territory factor. Plants are stationary, but even they compete over territory. Animals from insects to man are also territory conscious. Man builds fences around his territory or fights for his life over his country. A bird will fight off invaders of similar species to defend its territory. A domestic dog marks territory by urinating and is usually most upset if another canine invades it.

The territorial behaviour of the white or square-lipped rhinoceros is most interesting.

Firstly, as in most species but not all, the male of the species is territorial. In the white rhino, one territorial bull maintains a territory anything up to 250 hectares in size in the wild. There is no fixed shape to this area. The territorial bull will mark his territory by urinating backwards in a fine spray. All rhinos urinate backwards between their legs, but usually in a steady stream and not in a spray. The bull will paw the ground with his feet or scuffle the ground, bushes or bark of trees with his horn and will then proceed to urinate on the marks, so that his smell

On trail watching a territorial rhino bull

will impregnate the object. Any other bull or male of the same species then knows that that particular area is occupied. The rhino bull usually marks his territory from the centre outwards, like the spokes of a wheel, and this often leads to confusion as it tends to overlap another bull's territory.

The territorial rhino bull is the only male that will actually mate with the cows. Even if there are mature males within his territory, they will not mate until they have their own territory. A territorial bull will be challenged from time to time for the territory, and usually by a mature bull from outside his own territory. If the old territorial bull loses, he will then either stay where he is and adopt the attitude of a subordinate bull, or he will move out and find new territory. The cows are not territorial and will wander from territory to territory until held by a particular dominant bull when on heat.

THE MIDDEN

The white rhino has a habit of consolidating his toilet by defecating on the same spot in a particular area. This accumulation of dung is called a midden.

The territorial bull, when defecating on a midden, will once more show his dominance by scraping his dung over the rest of the midden, so that once again his smell is dominant. Middens with hollows in the middle are usually signs of a territorial bull's presence.

White rhino

The midden itself is a very interesting example of interacting cycles in nature. The fresh dung attracts many insects to it, for the purposes of breeding and food. The dung-beetle, for instance, will collect and manufacture balls of dung, which it rolls away and buries elsewhere. It is believed that these balls store the eggs of the beetle and of course will be the food supply for the new grubs when they hatch out. These balls are often rolled some considerable distance and then buried, thereby also scattering the dung and enriching the soil elsewhere.

Flies and other insects are also attracted to the midden, and these, too, lay their eggs in the dung. The insects and their larvae attract insectivorous birds; these birds again attract the smaller wild cats. Monitor lizards are also attracted to the middens for food. So a rather comprehensive natural cycle revolves around a midden.

The African Bushveld

If one were to travel with the navigator of a light aircraft and he was able to plot the latitudes of an area known as the African bushveld in the southern part of the continent, one would become forcibly aware of the brown, almost desperate dryness of the vast tracts of land, broken sharply here and there by green patches outlining tree-lined river-banks. Even the river-banks appear few and far between in the immense spaciousness of this dry, dusty land.

Within the bushveld latitudes, besides the infrequent patches of green, one would notice the areas of pale yellow desert, sparse in vegetation, as well as the flat, grass-covered plains of the savannah.

The vast belt of semi-arid bushveld country is further characterized by dense, grey thornbush, several types of acacia, primeval-looking baobab trees and the tall, sulky, yellow-barked fever trees which flank the bushveld river-beds.

In the dry season, one is struck by the wide river-beds, blessed only with a few muddy water-pools, and the dry silt and sand covered with a hundred different shapes and sizes of animal tracks and spoor.

Compared to central Africa, the southern bushveld region supports fewer species of plants, birds and insects, but the mammal life abounds there in great herds, both in number and type.

In the realms of the bushveld, the rhino and the baobab travel the same path, while balancing rocky outcrops harbour baboon troops, vultures and a leopard or two. One could sit in the scant shade provided by a giant baobab tree and identify many birds of the bushveld including the extroverted and colourful hornbill.

Baobab *Adansonia digitata*

The bushveld is indeed a grand mixture of animal life—grazing and browsing beasts, birds of prey, beetles, springhares and lizards, and the most masterly predators of all, the cat families.

Deep in this thick bush country, the carnivores and herbivores thrive together with one common aim—to survive—living out the dry seasons in a perfectly adjusted pattern, waiting patiently for the rains and the promise of greener pastures somewhere in this dry, stony land.

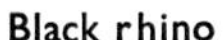

Black rhino

Animals of the Bushveld

The bushveld stretching from Moçambique in the east, north to the Zambezi, south to the eastern and northern Transvaal and Zululand, westwards across Botswana through to South West Africa, is inhabited by a widely diverse range of animals all linked to an intricate ecosystem, competing, adapting in a never-ending road of survival.

One sees the black rhino inhabiting dense bush shared by elephant and many different types of antelope; the white rhino preferring more open habitat as he is a grazer; the calf running in front of the mother whereas the black rhino calf follows behind, leaving the parent to penetrate the bush. Often one may wonder at the apparent lack of wildlife at a particular time of the day and later find the bush alive. Wildlife is constantly on the move, feeding, stopping to drink and resting in the heat of day.

Vultures

Frequenting the deep bush one finds nyala and the shy bushbuck along with numerous birds living in the undergrowth, the beautiful, though not often seen crested guinea fowl and many species of francolin who are the first to greet the dawn and the last to say goodbye.

The far-off moan of a well-fed lion mingles with the winging cry of the hadedas as they fly downstream with the early morning mist lifting. Soon

Lioness with cubs

the bush comes alive to the sounds of many other birds as well as other wildlife—a crashing sound at the river's edge as a waterbuck scents some unseen danger. The sun streams now through tall sycamore fig trees as it greets the day. The often presence of the mosquito is gone now until nightfall, and is replaced by different varieties of insects, seeking food and in turn being preyed upon by a host of smaller game and birds.

A family of mongoose runs back and forth as the distant boom of a ground hornbill floats by on the crisp morning air. Browsing kudu encounter a troop of baboons, neither seeking to harm the other, each concerned with the search for food. The sharp bark of a sentinel sends the kudu off, horns striking branches. A clatter of stones, and the troop seeks safety in the thorny height of protecting acacia trees. Numerous species of

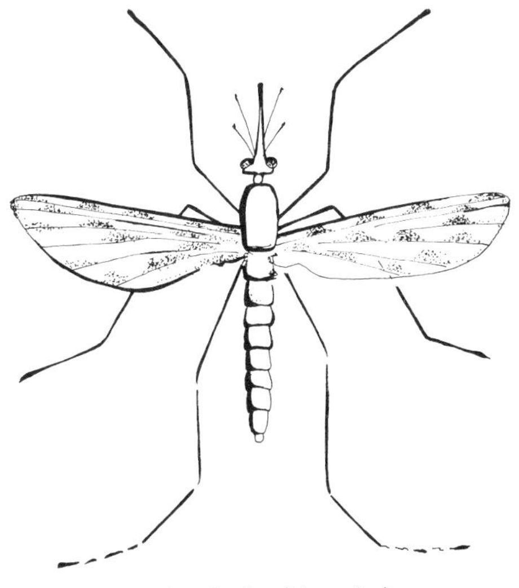

Anophaeles (Mosquito)

acacia are found throughout the bushveld, giving off that grey metallic look in the heat of noon. Small-leafed, the tree is well suited to arid, dry areas. Transpiration is very slight owing to the nature of the tiny leaves. Armed with a grey bark it resists the most intense heat, and its variety of thorns protect it from wildlife. It does, however, provide much food in the form of seeds, rich in protein. However, it does not escape giraffe or black rhino, who have no difficulty in coping with the tree's thorny defence. A pack of wild dog uttering their strange cry lope by, engines of death in carnival suits, jumping up on their hind legs to get a clear view across the tall grass. The bush is strangely quiet as this most feared of predators presses on in search of unseen quarry. The liquid rattle of a crested barbet calls from across the vlei and the sudden harsh cry of the 'go away bird' echoes from atop a giant baobab.

African buffalo (*Syncerus caffer*)

The elephant herd in single file disturb a crowned plover, uttering its shrill call as it spirals upwards. The elephants, pausing to test the wind, move on to water, mothers pressing little calves along with great patience.

Handsome, spectacular, the roan and sable prefer open woodland. These selective grazers, susceptible to habitat change, are attended by the ever-active oxpecker, who acts as honorary sentinel.

Vultures wheel in a cloudless sky, much a part of the intricate ecology of the bushveld.

The sun goes west on this African day as shades of night rapidly enfold all. A jackal calls, followed by another, to be overtaken by the laughing of a lone hyena. The sounds of the African bush are many as nocturnal creatures begin the long night.

The lilting ground call of the African nightjar, rivalling the fish eagle, as one of the best-loved sounds, is broken by the harsh scream of a galago, and the desperate call of the zebra as the lion makes his kill. Life moves on in this dry, thorny land.

Wildlife in Danger

The rapid deterioration of wildlife in Africa is merely one indication of the disregard man has for the environment. The term 'technosphere', which relates to the industrial environment that has evolved since the second century, could easily overwhelm the 'biosphere', the thin, protective layer surrounding the earth's atmosphere.

Cheetah skin

Considering endangered wildlife, we discover that hundreds of animal species, including many of the higher vertebrates, are in danger of global extinction. Through the course of evolution, extinction is a natural process and the fossil record which has been studied in depth for over a century provides some measurement of its natural rate. It is now accurate to assume that the current extinction rate of wild living things has been quadrupled by man since he imposed the technosphere on the planet.

Wattled crane — an endangered species

Looking back a few centuries we find that, since the year 1600, 359 species have disappeared, and today 922 species are classified as 'endangered'.

South Africa supports the greatest example of a Pleistocene type of fauna and flora, for example: 400 different kinds of mammals, of which 80 are considered spectacular; 850 species of birds, 425 of which have been recorded in the Ndumu Game Reserve alone. The latter figure represents 5 per cent of the total bird species population of the world. South Africa also supports approximately 18 000 plant species, 3 000 more than the United States, which is seven times the size of South Africa.

One of the animals which became extinct and which was indigenous to the southern part of South Africa was the quagga. The last known surviving quagga died in 1875 in the Berlin Zoological Gardens. In the wilds, it may have survived in the Orange Free State until as late as 1878. This animal was at one time so plentiful that the skins were utilized as grain-bags. In the end, the demand outweighed the supply.

Another animal—the bloubok—is not known to have survived beyond about 1800 and was apparently the first of the recent African mammals to become extinct. The bloubok was evidently restricted to parts of the south-west Cape zone. The tragedy of the bloubok is that its life as a species was of such short duration that it was hardly known to science before it disappeared.

The Cape lion was the first of the African subspecies of *Panthera leo* to become extinct. This animal was last recorded in the Cape in 1858. It was once common throughout the southern and central parts of South Africa. The huge black mane which completely covered the animal's shoulders, the thick growth of hair which traversed the latter end of the abdomen to roughly between the legs, and the broad, rather short head, distinguished the Cape lion from the present-day subspecies.

Specimens of the Cape lion are still preserved in the museums of Paris, Stuttgart, Leyden, Wiesbaden and the British Museum's store in Acton. Subspecies still found in South Africa include *Panthera leo krugeri*, confined to the Kruger National Park and its environs, and *Panthera leo vernayi* from South West Africa, Botswana and the Kalahari Gemsbok National Park.

Prior to the arrival of the European settlers, vast herds of grass-eating animals roamed the savannahs. But from the eighteenth century onwards their numbers were progressively decimated as a result of indiscriminate hunting. The bloubok and the quagga were wiped out, and only the creation of game reserves prevented such species as the bontebok, blesbok, black wildebeest, mountain zebra, black rhino, white rhino and the African elephant from suffering a similar fate.

According to the biologist, an endangered species is one that will become extinct if the trend of exploitation is continued. No one really knows the actual number of animals remaining, and counting the animals in the wild is a difficult task to say the least. A species is endangered when, as in the case of the cheetah, it has disappeared from areas where it formerly occurred. In India, for example, where cheetah once abounded, they are no longer there.

The African cheetah lives in open, semi-arid grasslands, bush or savannah. The lands which once belonged to the cheetah are now being occupied by commercial ranchers who are disturbed by predators. The lands are also being colonized by populations spilling over into the more fertile country. It is this competition for the land between man and nature, and, more important, the continual destruction of natural habitat, which present a pressing problem for the survival of the cheetah.

Cheetah skins hang like washing

The trouble is that the cheetah, like the fur seal, belongs to everybody and nobody. It is imperative that the cheetah be fully protected, and the continual fur trade in spotted cats be completely halted at every level in South Africa. As long as trade is permitted, this animal, along with the leopard, will be in danger.

Brown hyena—an endangered specie

Today, the brown hyena and the wild dog are two animals which are extremely rare in South Africa.

The future survival of the cheetah may depend on man's study of his ecological requirements and methods of safe translocation. The biology of the cheetah has been studied in fair detail and many attempts are being conducted on breeding them in captivity. But what do we know of his status in the wild. Outside of protected areas his chances of survival are slim. Racing to extinction, the cheetah is the fastest animal on earth. Long legged and lean, but cannot outrun civilization.

The Ocean

OCEAN CURRENTS AND THEIR EFFECTS

A warm, tropical current flows east to west across the Indian Ocean. On hitting Malagasy, it splits into two streams. One of these flows around the northern tip of Malagasy and then southward along the Mozambique coast as the Mozambique Current. The result of this is a tropical shoreline from Beira to northern Zululand. In the region of northern Zululand, this current is joined by that which flowed south of Malagasy, and the combined current is now called the Agulhas Current. This flows south along the coast of Natal and the Transkei towards Cape Agulhas, being forced ever seaward as the continental shelf becomes broader in the south. A cool inshore current flowing in a north-easterly direction penetrates between the Agulhas and the shoreline between Cape Auglhas and the Bashee River. The subtropical fauna of Natal is therefore replaced by a warm, temperate fauna along the Transkei to Cape Agulhas.

On the Atlantic side of South Africa, one finds the effects of the cold Benguela Current, flowing in a north-westerly direction. This current originated from subantarctic water. At approximately 40° South it sinks below the warm South Atlantic water and then travels north at a depth of between 200 and 500 metres. Southerly winds blowing offshore along the Namaqualand and South West African coast blow the surface water with it and the warm water is replaced by the cold subantarctic water.

This cold Benguela Current is rich in nutrients beneficial for the growth of a profusion of seaweeds, algae and phytoplankton, making the water look quite grey in colour in contrast to the clear blue tropical seas.

The seashore fauna along the west coast includes fewer species in comparison to the east coast temperate and tropical faunas, but there are greater quantities and larger individuals of each species.

WAVE ACTION ON OUR SHORES

As a wave breaks, the surface water is carried forward and the bottom water flows back as undertow, thus the fine sand at the bottom is carried away to be deposited during calmer times. Coarse sand on beaches is usually an indication of heavy surf. Rocks on ledges exposed to this sand-paper action of sand in water are continually being polished and worn away, and are fairly free of barnacles and seaweeds. In sheltered bays

Wave action

and estuaries, where there is little wave action, seaweed and grasses are more prolific and support many forms of life. In strong surf conditions, only the most firmly attached seaweeds, red-baits, barnacles and limpets can support themselves, while the more delicate forms are torn off or are restricted to sheltered crevices. Some animals depend on the water movement to supply food or oxygen; the barnacle Octomeris is a good example of this. The wash and spray from waves carries water well above the level of the sea and on exposed headlands the marine fauna may extend to some 6 metres above sea-level, while in sheltered areas the top of the marine fauna corresponds almost exactly with the level of the high tide.

Wave action on our coastline

TIDES

Tides are caused by the attraction of the sun and the moon upon the water of the rotating earth. Obviously the moon has a greater effect, as it is closer, and as it passes over the ocean basins it sets up a motion similar to water in a basin washing up and down the sides as the basin is tilted from side to side.

In most oceans there are two periods of high tide and two periods of low tide per day, but as the moon rises 50 minutes later each day, the two periods total 24 hours and 50 minutes. If low tide is, say, at 7 a.m. on Monday, it will be low tide again at about 7.25 p.m. that night and about 7.50 a.m. on the Tuesday morning.

When the sun and the moon are pulling in the same direction (e.g. when there is no visible moon), or in diametrically opposite directions (at full moon), their combined effects on the tides are great. This is when spring tides occur, when we have the highest high tides and lowest low tides. During the periods of first quarter and last quarter of the moon, that is when the sun and moon are pulling at right angles to one another, there is little difference between high and low tide. This is called a neap tide.

The tides, therefore, are related to the phases of the moon, and the approximate time of the tides can be forecast if we know when the moon will be full.

Full Moon—Spring tides with a range of 1,8 to 2,1 metres. Low water springs (L.W.S.) about 9.00 a.m. or 9.25 p.m. and high water springs (H.W.S.) over six hours later.

Last Quarter—Neap tides with a range of 0,3 to 0,6 metres. Low water neaps (L.W.N.) at about 2.50 p.m. and 3.15 a.m. High-water-neaps (H.W.N.) over 6 hours later.

New Moon—Spring tides—about the same times as for full moon.

First Quarter—Neap tides—about the same times as for last quarter.

ZONES

The intertidal zone is an area between high and low water marks. Here in this zone the intertidal fauna and flora are very much affected by the rise and fall of the tides. At low tides, animals and plants are subjected to heat and light from the sun, and also the drying action of wind. On damp and cloudy days, the conditions will not be so severe, but during heavy rainfall periods the flora and fauna are subjected to extreme changes in salinity. Many sea animals cannot feed unless they are covered by sea water.

Life on or near the high water mark is more seriously affected than that at the low water mark. For instance, animals and plants living at high water springs may go for about a week without being covered by salt water. Therefore, life at H.W.S. is hardier but obviously scarcer than that at low water springs.

The area between high water springs and low water springs can be visibly graded into bands according to tolerance to conditions. The banding or zonation is most easily seen by noting the dominant animals and plants at various tide levels. Above high water neap, the rock is likely to have periwinkles of the genus Littorina as the dominant species. Thus it is called the littorina zone. Barnacles dominate the middle of the shore, mostly the genus Balanus, and is known as the balanoid zone. The upper half of this zone has little seaweed in comparison to the lower half, so here we can distinguish an upper and a lower balanoid zone.

Along the Cape shores, the zone below the lower balanoid zone is dominated by a pear-shaped limpet known as *Patella cochlear*; this we call the cochlear zone. Along Natal shores, this cochlear zone is absent and is replaced by a carpet of algae.

Finally, the lowest levels of the shore seen between the waves are dominated along the Natal–Cape coast by red-baits (Pyura) and a large kelp (Ecklonia). This marks the upper fringe of the fauna below the intertidal zone and is known as the infratidal fringe.

SALINITY

Sea water is made up of many different salts that have been washed down the rivers since the dawn of time. Most of the salt is sodium chloride, but potassium, calcium, magnesium, sulphate, borate, carbonate and bicarbonate are also contained. Several of these are important to the growth of marine plants. Those of some importance are nitrates, phosphates and manganese. Salinity is expressed in parts per thousand and the salinity of sea water is 35 parts per thousand, i.e. 35 per cent salt content over a thousand.

Marine life is adapted to a certain range in salinity, but most of it will die if the salinity levels double or halve. Therefore it is found that very little life is found in rock-pools where there is a great amount of evaporation thereby raising the salinity drastically. Brackish water has the same effect on marine life.

ESTUARIES AND THEIR IMPORTANCE

An estuary may be considered to be where one or more rivers converge to flow into the sea, often forming lakes or lagoons.

Fresh water flowing down the river mixes with sea water brought in with tides and the salinity obviously varies from 0 to 35 parts per thousand, i.e. from the head of the estuary to the mouth. Salinity will of course change according to the seasons. For instance, the rainy seasons will drop the salinity level and dry, windy seasons will evaporate and cause higher salinities.

The depletion of fresh-water supply to our famous St Lucia Lake, which is the largest estuary in the Republic, has caused such extremes in salinity that it is sometimes feared that the whole system will collapse. In fact, recently, as a result of some 56-odd rivers and streams that run into this lake having being diverged or dammed, the salinity level at the head of the lake rose to three times that of sea water and the lake virtually died. For the first time in living memory crocodiles died through dehydration, fish died, and in fact the food-chain collapsed, causing the water-birds to leave in their thousands.

Water-birds left in their thousands

The fish which inhabit our South African shoreline do so according to the availability of food. Thus fish for the most part inhabit the areas of the continental shelf, fairly close to the shoreline, and their movement is one more parallel to the coast in search of food than that of swimming out into deep sea and back.

Fish spawn at the mouth of estuaries, and it is from here that the young fish swim up into estuaries in search of food and protection. The edges of or marginal shores of natural estuaries are covered in rich feeding-grounds of Zostera (eel-grass) and Ripia. These shallow feeding grounds are the basis of the food-chain in estuary life and it is here that most life teems and abounds, from the smallest microscopic animal and plant life to crustacea, fish, birds and, naturally where they occur, crocodiles also.

Islands within estuaries are important, for they allow a larger area of shallow feeding-grounds around their fringes.

When the fish reach maturity, they swim out to sea again and the whole cycle is repeated.

Hippos are tremendously important to river and estuary life. They are nature's natural dredgers, and in rivers which tend to silt up they displace the sand or silt towards the banks and cause a continual flow of water. The absence of hippo in most of South Africa's rivers have caused them either to silt up altogether or to form vast sandbars at river mouths, creating lagoons, into which it is difficult for the young fish to enter unless there has been extensive flooding. Hippos also create paths from marginal swamps into rivers and estuaries, causing water to flow from underground water-tables into the estuaries. They also have the habit of defecating in the water, spreading the dung with their tails. The dung, which is pure grass, is fed upon by fish. Thus hippos are most vital to estuarine systems. It is rather tragic that most of the estuaries in South Africa are now devoid of hippos and in fact of most life.

Pollution is another cause of disturbance in estuaries. Some of our South African rivers are so badly polluted from factory effluent that they are completely dead. Some in fact are so polluted that life out to sea in the area of the river mouth has also been wiped out.

See overleaf for captions

Photo: Don Richards

Previous page
46. Down-graded habitat caused by overgrazing

47. Donga erosion

Above
48. Waterhole turned to mud by elephant

Opposite
49. Amphibious frogs and salamanders are members of the littoral zone

50. Pans on the eastern shores of Lake St. Lucia

51. Pond in the Sabie-Sand Wildtuin

Following page
52. Fire—a useful tool in land management

53. Oil pollution—death for thousands of sea birds

▲49
▼50

▶51

▲52 Photo: Brian Kemp

Photo: A. Gifford ▼

The Fresh-water Environment

In our part of the world, where drought is a common condition, the presence of rivers, streams, lakes and pans are like oases in the middle of a desert.

Trekking along a dusty wilderness trail, one is constantly aware of the absence of water—a valuable commodity in our times—and there is always a sense of relief in sighting a river-bed, a swamp or a hidden stream.

Apart from the aesthetic value of rivers and lakes, and the simple need to drink water, there are millions of life forms contained in fresh-water habitats which are constantly changing with the effects of the seasons, rain and sun, and erosion, which often results in water habitats turning into terrestrial habitats.

Fresh-water habitats provide fascinating opportunities for the study of large colonies of animals living in the zones of lakes, ponds and swamps, or streams, rivers and spruits.

Fresh-water environment

One zone might support many rooted plants with floating leaves like the water-lilies, while another zone carries submergent vegetation with rooted plants living in a constantly submerged state.

Fresh-water environments are the homes of animals ranging from snails, dragonfly nymphs, water-beetles and bugs to frogs, turtles, water-snakes and the many species of fish such as rainbow trout and tilapia.

Silting of the White Umfolozi River

Fresh-water habitats may be divided into two categories, namely—

Standing-water, or lentic (levis, calm), habitats:

Lake—Pond or Pan—Swamp or Vlei.

Running-water, or lotic (lotus, washed), habitats:

Spring or Spruit—Stream—River.

As in land habitats, these fresh-water habitats are prone to change and one should never consider what one sees now as the norm. For example, erosion from land and the activities of plants tend to eventually fill up pans and lakes, possibly producing terrestrial (land) habitats.

Streams and rivers tend to cut down to base level and change as a result of the action of the water. When base level is reached, the current is reduced, sedimentation takes place, and a shallow, meandering river results. This is sometimes known as an 'old river'. This process can be speeded up considerably by man's abuse of soil such as ploughing down to the river's edge and overgrazing in catchment areas, a situation which is only too common in South Africa. The Orange River alone washes down some 400 million tons of top soil in a year. The Vaal River takes away the equivalent of a 16 hectare farm a year. The Tugela is a good example of a country losing its life-producing soil. Many of South Africa's rivers have become so silted that they dry up completely for nine months of the year and become virtually storm-water drains. Naturally, if one dug down, water would be found, and this is exactly what a great number of people and animals have to do.

The amount of silt in any fresh-water system will affect the penetration of light, restricting the process of photosynthesis.

In lakes and pans, three zones are evident, each with its own type of life. These are:

Littoral Zone—the shallow-water region with light penetration to the bottom, occupied by rooted plants.

Limnetic Zone—the open-water zone to the depth of effective light penetration. The community in this zone is composed only of plankton and other forms of microscopic life.

Profundal Zone—the bottom and deep-water area which is beyond the effect of light penetration.

In streams and rivers, two zones are evident:

Rapids Zone—shallow water where the current is great enough to keep the bottom clear of silt, therefore producing a firm bottom. This zone is occupied by life forms which have adapted to 'cling'.

Pool Zone—deeper water where the velocity of current is reduced, resulting in a soft, silted bottom, favourable for burrowing forms of life.

The littoral zone, because of its importance, is described further, as follows:

Within the littoral zone, there are three subzones, namely:

(*a*) Emergent vegetation—rooted plants with leaves above the water. The emergent plants, together with those of the moist shore, provide an important link between land and water. They are used for food and shelter by amphibious animals and provide an easy means of entry into and exit from the water for aquatic insects which spend part of their lives in the water and part on land.

(*b*) Zone of rooted plants with floating leaves, for example, water-lilies. The undersurfaces of the lily pads provide resting-places for animals and a place for attaching eggs.

(*c*) Zone of submergent vegetation—rooted or fixed plants completely submerged. Leaves here are thin.

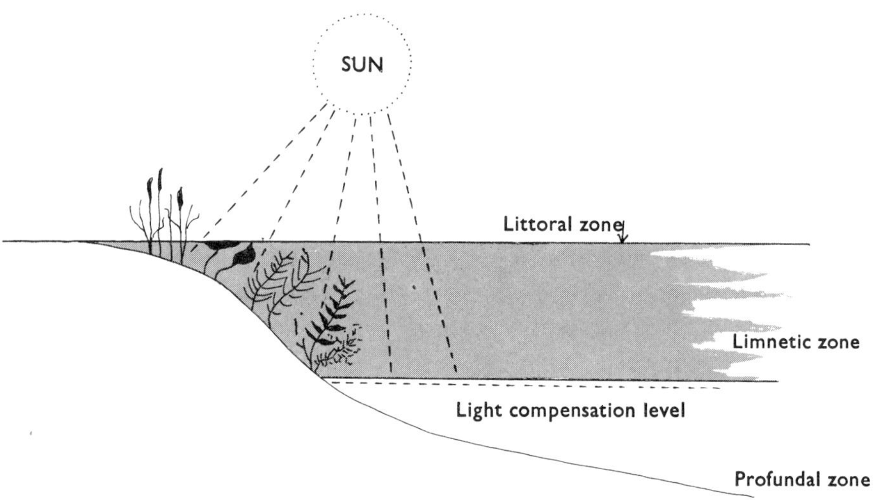

Lake and pan zones

The littoral zone in fact is the home of a greater variety of animals than the other zones. Animals ranging from snails, dragonfly nymphs, flatworms, hydra, to midge larvae rest on or attach themselves to stems and leaves. Snails feed on the plants, but most of the other forms mentioned are carnivorous. Water-beetles and bugs abound in different forms.

Amphibious frogs, salamanders, turtles and water-snakes are almost exclusively members of the littoral zone. Lake and pond fish move freely between the littoral and limnetic zones, but most species spend most of their time in the littoral zone; many species actually establish territories and breed there. The non-rooted producers of the littoral zone are comprised mostly of species of algae. Temporary rivers and pans, which are dry for part of the year, are especially interesting and support a unique community which must be able to survive in a dormant stage during dry periods, or they must be able to move in and out of these areas, as can amphibians and adult aquatic insects.

The eggs of some of these adaptable animals are able to survive in the dry soil for many months, and development and reproduction occur in a short time while water is present. This may even be seen in temporary pools on tops of mountain ranges. Some reptiles, such as the terrapin, are able to go into hibernation by burrowing into the mud. The temporary pan or pool or river is a favourable place for those organisms adapted to it, as competition is thereby much reduced.

Not all rivers are alike; some are swift-flowing and clear, others sluggish and brown in colour. Obviously, the nearer the source, the cleaner the river. Streams and rivers among the mountains and near their catchments usually flow over a rocky bottom. Here, the champagne-like water is the home of fish such as the rainbow and brown trout; small-mouthed bass also occur here.

Most Natal and Transvaal rivers become at some stage or another

Trout taking dragonfly

murky and chocolate in colour because of siltation. This is due to the nature of the land, which is easily erodable, but, as has been mentioned before, this erosion has in many instances been accentuated by man's mismanagement to cause shallow, sluggish rivers. Where rivers were V-shaped and rock-bottomed, they have now become U-shaped and have a sandy bottom. This leads to quick evaporation and drying up.

The rivers of the Cape Province are in most instances brown in colour but clear. This is not silt, but due to the mineral content of the rock the water passes through and over. The western and parts of the eastern Cape are mountainous and because of the rocky nature of the terrain there is less erosion and rivers are cleaner. Fish such as the barbel, mud-fish and kurper frequent the dirtier waters.

In Zululand from the Pongola River north to Mozambique and Rhodesia, many of the rivers have tiger fish, the sportsman's favourite fresh-water game-fish. Bream, or tilapia, frequent many rivers and lakes of southern Africa.

Many dams and streams in the Republic and Rhodesia have been stocked with fish such as trout, large-mouthed black bass, tilapia and blue-gill. By such stocking, farmers and parks departments are ensuring that sporting fish are available and, what is more important, it brings back a part of necessary life to the fresh-water environment.

Fire

FIRE AND ITS IMPORTANCE TO THE NATURAL VEGETATION

Fire is a natural phenomenon. Lightning fires have, in the natural world, occurred with regularity. In southern Africa, and especially in the grassy, savannah areas, certain adaptations have taken place, enabling plants to withstand the ravages of fire.

The grass of the savannah is normally tall and burns easily, especially when dry. Hence the trees have adapted accordingly by being umbrella-shaped, i.e. a canopy-shaped tree with few under-branches, enabling fire to sweep through without affecting the tree adversely. The barks of these trees are also normally thick and able to withstand fire. The bases of our Africa grasses are usually protected with a sheath, enabling the plant to withstand fire and giving the grass a chance to regrow.

Fire is used as a management tool on farms and in game reserves. Often the use of fire is abused and this can lead to severe depletion of soil cover and erosion. When used wisely, fire is a valuable ally of man.

Burning in game reserves, for instance, is done only for the following reasons:

(1) To burn the accumulation of litter which stifles the younger grasses.

(2) To attract animals off overgrazed areas on to the new burns.

(3) To supply fresh grazing.

One particular area may be burned only once in seven years and is burned always after the first spring rains, ensuring that there is damp soil and that the root systems are not destroyed. Large areas at a time are burned, for if the area was small it would soon attract vast herds of animals and the area would very rapidly become overgrazed and ruined.

Certain shrubs in our veld produce large quantities of hard-coated seed that tend to lie dormant in the soil until the scrub in which they occur is burned. Fire cracks the seed-coats and then seedlings appear. Fire-tolerant species of plants will normally increase in abundance at the expense of those plants that are sensitive to fire, often resulting in a dominant plant.

Some effects of burning are listed here:

(1) Increase in light at the ground surface.

(2) Lack of shade also allows the soil to heat up and cool off to a greater extent, so that the daily range of temperature is increased. Ordinarily, this results in a decidedly early development of vegetation in the spring.

(3) Burning of surface organic matter coupled with the effects of beating raindrops will increase run-off and rapid erosion.

(4) If fires are very hot and much of the humus is destroyed, the soil is subsequently impoverished.

(5) If fires are not very hot, their effect is to increase soil fertility and tends to raise the PH factor of soils, which tends to favour bacterial action and improves soil conditions for plant growth.

(6) Of great importance is the change in animal life brought about by fire. New conditions favour new kinds of animals, having different requirements for food and cover. Birds that prefer open vegetation congregate on burned areas and carry in seeds from plants growing on other burned areas that they have visited. Thus dense growths of berry-producing scrub may develop rapidly.

(7) Veld burning suppresses species of woody plants in the sense that it hampers the growth of seedlings, kills off aerial sections of trees and plants smaller than 2 m in height. It reduces bush encroachment. Established trees and shrubs are rarely totally destroyed by veld burning.

Forest fire

54. The green mamba. Highly poisonous

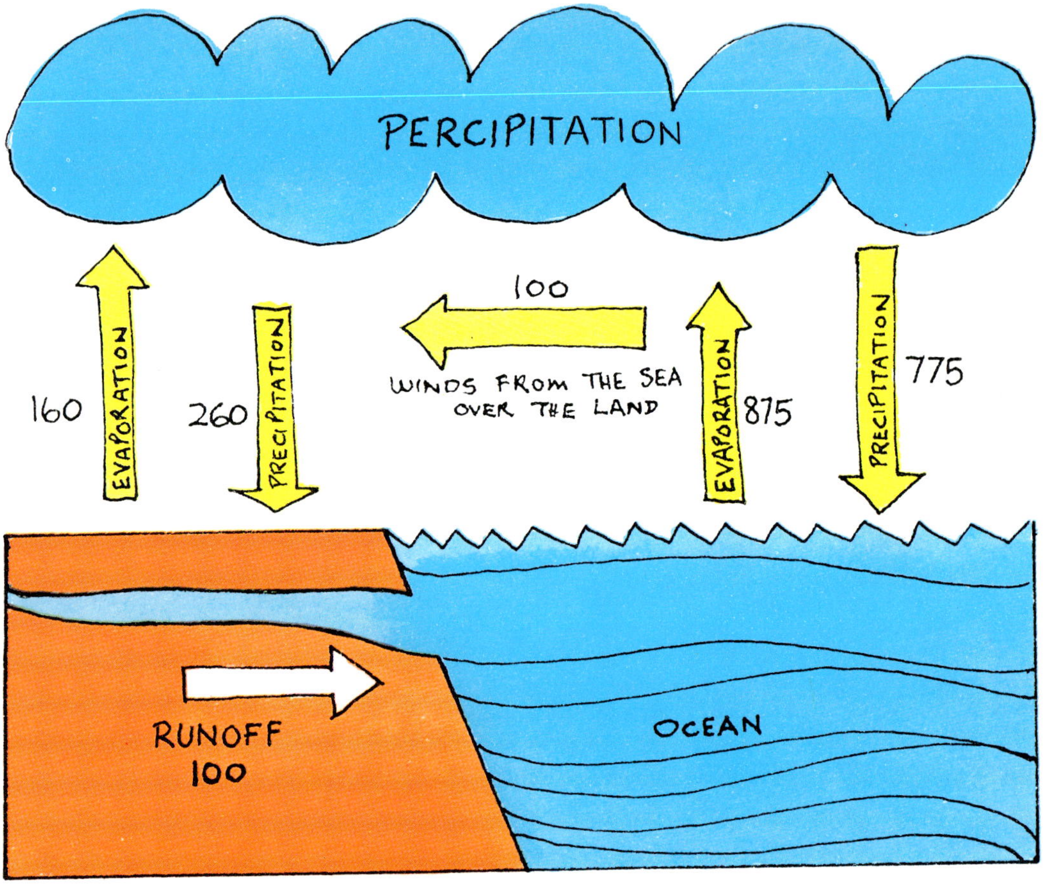

PERCIPITATION
EVAPORATION
160
PRECIPITATION
260
100
WINDS FROM THE SEA OVER THE LAND
EVAPORATION
875
PRECIPITATION
775
RUNOFF 100
OCEAN

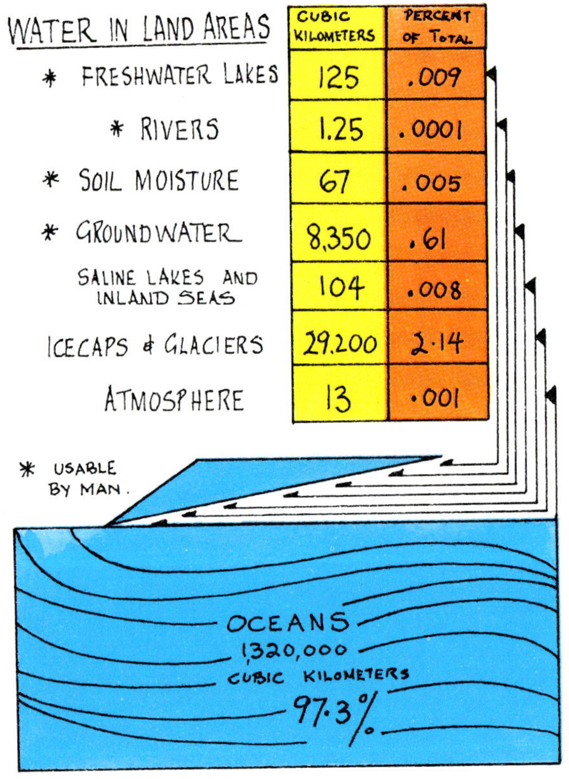

WATER IN LAND AREAS
CUBIC KILOMETERS
PERCENT OF TOTAL
* FRESHWATER LAKES 125 .009
* RIVERS 1.25 .0001
* SOIL MOISTURE 67 .005
* GROUNDWATER 8,350 .61
SALINE LAKES AND INLAND SEAS 104 .008
ICECAPS & GLACIERS 29,200 2·14
ATMOSPHERE 13 ·001
* USABLE BY MAN.
OCEANS
1,320,000
CUBIC KILOMETERS
97·3%

55. The Hydrological Cycle

56. The earth's fresh water

Man and the Changing World

Man is only one of three million species that inhabit this earth, but he already consumes more food than all land mammals put together. During the Stone Age period, the world's human population consisted of some one million people. It took until 1850 to reach one thousand million. Only 80 years later, in 1930, the world's human population doubled itself to two thousand million. In 1960 it was estimated at three thousand million

Oil pollution

and in 1975 it is expected to reach four thousand million, with the probability of eight thousand million by the year 2000. From the outset, man was more intelligent than other living creatures and was therefore able successfully to adapt and overcome the difficulties confronting him. During the Ice Age, man covered himself in skins instead of migrating to warmer climes. He then discovered fire and began to build shelters. From eating fruits, berries and roots, he learned to grow crops and to rear animals for meat, milk and skins. He started to trade, which led to road systems and eventually to the building of ships which connected him with other continents.

With each progression, man has advanced in all fields of life. Industry and scientific advancement have swept ahead without man learning to manage his world and its resources more rationally.

Rivers and lakes have silted up through soil abuse and incorrect farming methods. Usable water is becoming scarcer each day due to silt and pollution: only 1 per cent of the earth's water supply is now of use to man.

Atomic power can be very useful to man, but already man has polluted the earth with radio-active dust. Unwise use of poisoning insecticides has resulted in poisoning plants, animals and human beings, too. Factories have spilled their effluents (waste products) into rivers, lakes and seas, until they can take no more. We burn coal, paraffin and other fuels which not only pollute the atmosphere but take a long time to replace.

In the early days of the earth, people such as the Red Indians and the Bushmen lived in harmony with their environment; they knew that their lives depended on knowledge of sound ecological concepts. Today, we are only beginning to learn. We cannot treat our planet as if it were limitless in its supply. As we have learned earlier in this book, all source of energy comes to us from the sun via plants. The coal and fuels we burn, the petrol we use, have this energy. We are using these, our natural resources, as if they would never run out, but already we have over-used some of our sources of energy which took millions of years to form.

BEFORE IT IS TOO LATE

The litter problem in America today is enormous. 600 000 tons of litter have to be disposed of daily. This involves a tremendous amount of money; also, a large amount of this litter is not recycled and is therefore absolute waste. We have our own litter problems here, but as charity begins at home, so must we, in our own minds, be pollution and litter conscious. Plastics and bottles, papers and tins, are left by many careless people in our beauty spots, whether it be in mountains or in water. Apart from spoiling our beautiful heritage, these things can become a source of danger.

So litter, water, soil, food and our needs become a personal responsibility, where, with self-discipline, we must learn not only to respect ourselves but our environment, too.

Litter—the product of man

There is only one earth and it is not very large; the biosphere is even smaller, and what nature has created and developed over millions of years, and which allows our planet to teem with life for our wise use, is rapidly being shattered to such a degree that if we do not do something NOW, it will be too late.

INSTABILITY OF VEGETATION DUE TO MAN'S MISMANAGEMENT

Soil differences cause variation in vegetation. For instance, depressions will collect silt, and usually have a more palatable and greener vegetation than that of the surrounding veld. Soft, sweet vegetation under trees is due to rich soil caused by humus, shade protection and animal droppings, as those animals seek shade. Heavier soils around ant-heaps will carry better vegetation than the surrounding veld. Consequently, the concentration of animals on these above-mentioned areas is heavy, causing grazing pressure, denuded soil and conditions suitable for increased run-off of water. Thus, this selective grazing gives rise to increased areas of soil erosion.

The need for water is vital

Other examples of this phenomenon are evident in areas where livestock has been driven for long distances to water and kraaled in the same area every night.

Heavy selective grazing often removes climax species of grassland, while heavy, non-selective grazing preserves the climax species. This illustrates a principle that grazing should be heavy for limited periods, must not be continuous, and must be alternated with periods of rest.

Selective grazing has led to the almost complete disappearance of grass from the Karoo, and the establishment of useless veld types (a desert). Bare areas, do not normally occur naturally. Nature is always ready to heal unless man intervenes. There is always a tendency for natural vegetation to migrate from climax areas to desert conditions and so heal the wound.

On the advent of the European in South Africa, the population of game must have by far exceeded the present population of domestic livestock, and yet the vegetation was far richer than it is today. Wild animals have varying feeding habits and levels of feeding. They were always on the move, looking for water, giving the veld an opportunity to recover naturally. The habit of large 'trekking' herds during migration grazed the veld heavily but not continuously in the same place.

Our game reserves of today have, of necessity, to be fenced in, restricting animals to a local area, thus this area has to be scientifically managed according to its carrying capacity, and the game culled accordingly, especially if there are not enough prey species. Thus we have heavy selective grazing in some areas, leading to erosion. For example, old African kraal-sites in some game reserves appear to have produced climax grass patches in bushveld areas. These have been sat on by animals, reducing them to continual short cover, but always green and thus returned to by the animals.

Destruction of habitat by elephant due directly to unnatural factors by man

These areas are favourite spots of the white rhino, or selective short-grass grazer. Wallows have appeared in most of these old kraal-sites, due to water collecting from a fast run-off on almost bare soil. The wallow grows in size each year as more and more mud is removed by the wallowing rhino and warthog. Eventually, the wallow becomes a sizeable drinking-hole in the wet season, attracting more and more game. The surrounding area becomes denuded of cover through hoof trampling and sheet erosion sets in, often followed by donga erosion. This situation leads to increased run-off of water which should be going into the soil to revive the veld, and siltation of rivers, which again leads to loss of available water to animals and man.

Rhino Wallow

Snake Acrobatics

Snakes seem to be the subject of many myths and misunderstandings, for a snake's existence is one of the most striking adaptations in nature.

It is a myth that snakes are slimy creatures. On the contrary, their skins are dry and exceptionally clean. Ecologists estimate that there are no less than 2 500 snake types, silently slithering their way round the world, and South Africa supports many of these species.

The truth about snakes is more incredible than any of the fables one grows up with—they are never charmed by a snake-charmer's music. Snakes are deaf.

There is consolation in the fact that nine-tenths of the world's snake population are harmless and few are aggressive towards man. The dreaded venom helps in the capture of prey. A venomous snake rarely uses its poison against large mammals or human beings and then only in self-defence. A snake will, however, deliver a nasty bite if provoked. Fortunately, most snakes have long-range powers of perception which indicate the approach of man or animal. A snake has a built-in early warning system thanks to its highly sensitive underside which senses vibrations in its path. The adder family is notoriously lazy and does not react to vibrations as quickly as other types.

Everything about the life style of a snake revolves round the fact that it has no arms and legs. Unlike a simple worm, it is a complex creature with highly developed internal systems such as respiration.

The whole anatomy of a snake radiates sensory cells, providing it with the scent and feel of its immediate environs. The muscle structure is powerful and a supple backbone ensures that its essential gliding ability functions expertly.

A snake never closes its eyes because it does not have eyelids. No snake has a poisonous sting in its tail and its fearful-looking tongue is harmless. Killing a snake will not invoke its mate to seek revenge.

Snakes live in a peaceful world of silence, which is the way they prefer to live. They fit neatly and deftly into one of the many complex food-chains, performing some of nature's vital functions.

Man does not form part of a snake's natural existence. Snakes do not need man, but should man need treatment from an inadvertent bite, here is a comprehensive anti-snake-bite chart.

TREATMENT OF SNAKE-BITE

N.B.: First-aiders should be familiar with the general notes *before* going into the field.

GENERAL NOTES

Many, if not most, snakes are not dangerous.

Victims usually need reassurance and should be moved only after proper assessment.

Cobra and adder bites are often non-fatal, and treatment can be withheld until symptoms appear if there is doubt about identification. Also, the snake does not inject venom every time it bites.

If first-aiders do not know how to inject intravenously, drugs can be given intramuscularly. *But N.B.:* Snake-bite anti-venine is highly effective by intravenous route even after a long delay; so if you cannot get into a vein, use a portion only intramuscularly, then take the patient to a doctor or hospital. Injecting part of the antiserum around the area of the bite may be of value if done early (within 5–15 minutes) and with care not to aggravate local swelling. It will not eliminate necrosis.

The other drugs used for resuscitation work well by intramuscular route, so do not hesitate to use them, if required, before moving the victim.

Cortisone and antihistamines are most valuable both in the treatment of poisoning *and* in the treatment of sensitivity to antiserum. It is recommended that antihistamine be given in all cases.

Opiates (morphia, omnopon) and barbiturates (sleeping drugs) depress respiration and should not be given except by a doctor.

Children require *full* doses of antisera, but doses of other drugs are calculated according to the formula: *Age in years*/15. *N.B.:* All antisera are combined in one ampoule but treatment is described separately for different snakes to assist in description of symptoms.

Antibiotics: Infection often supervenes on snake-bites, so antibiotics should be administered after the emergency treatment is completed.

Tetanus is uncommon, but a booster of toxoid is worth giving.

SUGGESTED CONTENT OF KIT

1. Two 10 ml disposable syringes and two 21-gauge needles.
2. Two 5 ml disposable syringes and two 22-gauge needles.
3. Two 2 ml disposable syringes and two 23-gauge needles.
4. Four ampoules of polyvalent antiserum type—effective against adder, cobra, mamba and rinkhals.

5. Cortisone—6 ampoules of hydrocortisone 100 mg, or the equivalent, for intravenous use.
6. Antihistamine—4 ampoules for intravenous use.
*7. Pethidine (for pain)—2 ampoules of 100 mg each.
*8. Adrenaline (1 in 1 000 solution)—5 ampoules.
9. Cotton wool and skin antiseptic, e.g. meths.
10. Eyebath and plastic water-container for rinsing eyes (spitting cobras).
*11. Injectable antibiotics.

To be used only by a doctor.

Cobra and Mamba Bites

(Front fixed-fang or Elapid snakes)

The majority of these bites will be due to cobras. This group includes rinkhals. Poisoning is rapid and mainly neurotoxic:

Symptoms: Disturbed vision, drooping eyelids, slurred speech, difficulty in breathing. Later, coma and paralysis.

MAMBAS especially will produce symptoms within 10 to 20 minutes; often with pain in throat and difficulty in swallowing. Severity and speed of onset are diagnostic of mamba bites. If mamba identification is positive, start treatment at once, without trial test.

If cobra bite suspected, time permitting, test for antiserum sensitivity and wait for symptoms to appear:

Inject one drop of polyvalent antiserum into skin of forearm and watch for 10–15 minutes. If victim *is* sensitive, an ugly red weal 12–25 mm in diameter will form round the injection.

While you are waiting, inject 2 ml antihistamine intramuscularly (upper outer quadrant of buttock).

TREATMENT

1. Do not move patient.
2. First inject 2 ml antihistamine, intravenously if possible (or else into muscle of upper outer buttock).
3. Inject 2 ampoules of cortisone.
4. Then inject 20 ml (2 ampoules) antiserum intravenously.

   If first-aider unable to use intravenous route, inject intramuscularly, and prepare for (*a*) resuscitation, (*b*) moving to a doctor.

If patient *is sensitive* to the antiserum, shock will develop rapidly: cold, clammy, sweaty, fast faint pulse, maybe loss of consciousness.

Treat SHOCK as follows:

1. 2 ampoules of cortisone, preferably intravenously. Another 2–4 ampoules can be given within 5 minutes if no response.
2. Another 2 ml antihistamine (preferably intravenously).
3. $\frac{1}{2}$ ml adrenaline, intramuscularly for acute shock; this can be repeated every 10–15 minutes but only if condition remains critical.
4. 100 mg pethidine will relieve pain and calm the patient.

If victim responds well to this treatment, more antiserum can be given for persistent or recurrent symptoms. You *must* give more antiserum for all mamba bites.

ADDERS

(Erectile-fanged or front-hinged fang)

Short, fat, stumpy-tailed snakes which hiss, coil low and seldom bite higher than the ankle.

SYMPTOMS

Pain and swelling at site of bite are severe. General symptoms are slow in onset:

Cytotoxic (local tissue destruction) and haemotoxic (persistent bleeding from fang-marks, bleeding gums, bruises all over body).

May later develop shock: dizziness, vomiting, perspiration (cold, clammy patient), fast and thready pulse, collapse.

TREATMENT

No tourniquets.

No local cuts.

No local antiseptics.

Because onset of symptoms is slow, take time to *test for sensitivity to antiserum*:

Inject one drop of polyvalent antiserum into skin of forearm and watch for 10–15 minutes. If victim IS sensitive, an ugly red weal 12–25 mm in diameter will form round the injection. While you are waiting, inject 2 ml antihistamine intramuscularly (upper outer quadrant of buttock).

No sensitivity

If there is no reaction, or only a mild pink discoloration, proceed as follows:

1. Inject 10 ml (1 ampoule) of polyvalent antiserum in amounts of 1–2 ml around the bite but NOT into swollen parts. This is only important if done soon after the bite.

2. Inject 20 ml (2 ampoules) intravenously. If you do not know how, inject half this amount deep into the muscle of buttock (upper outer quadrant) and arrange to move the victim to a doctor or hospital.

If the victim's condition deteriorates gradually over 30 minutes or longer, inject more antiserum and use antihistamine.

Is sensitive

1. Inject 2 ml antihistamine intravenously.
2. Wait for symptoms of snake venom poisoning to appear. Transport to doctor if possible.
3. Inject 10 ml (1 ampoule) intravenously if symptoms become severe.
4. Treat for sensitivity as under mambas.

Gaboon Adder

Highly poisonous, so start treatment at once—no trial dose.

TREATMENT

1. Inject antihistamine intravenously.
2. 30–40 ml (3–4 ampoules) of antiserum intravenously. If unable to inject into a vein, put half dose into muscle of upper outer buttock and rush to a doctor.
3. Treat sensitivity with cortisone etc. as described under adder bites.

Berg Adder

Poisoning by venom is not responsive to antiserum, so mainstay of treatment is cortisone. Inject 2 ampoules immediately, another 2 if symptoms develop. Preferably intravenously, otherwise intramuscularly. Supplement this with 2 ml antihistamine.

Burrowing Adder

Not highly poisonous (usually) but often not recognized because the snake looks different from usual. (Thin and dark.) Often mistaken for the harmless wolf snake.

Boomslang

Bites are rare, because fangs are at back of mouth. Poison is haemotoxic—interference with blood clotting. Serum is not commercially available, but urgent telex or telephone call to South African Institute for Medical Research, *Johannesburg* 724-1781, should be made and serum will be sent by fastest route.

Symptoms are slow in onset, and intravenous use of serum even after 2–3 days is still highly effective.

MEANWHILE

1. Take victim to hospital and prepare for availability of blood transfusion.
2. Proceed with general supportive measures: pethidine 100 mg for pain; cortisone 2–10 ampoules intravenously or intramuscularly as indicated. Antihistamine 2–4 ml intravenously adrenaline ½ ml intramuscularly for acute collapse (unlikely to happen).

Spitting Elapids

If victim is seen immediately or shortly after incident, wash eyes copiously with water. No further treatment is required unless there has been a delay, in which case add one part of polyvalent antiserum to ten parts water for wash-out.

Milk is NOT better than water, but can be used instead (or any bland fluid i.e. 'coke', beer.)

Puff-adder—belonging to the erectile-fanged group. The illustration below shows the fangs in the resting position. The fangs are rotated through 90°.

Puffadder

Other members of this group are: sidewinding adder, many-horned adder, berg adder, gaboon adder.

Mamba—belonging to the front-fanged group. Penetration of the fangs is affected by a biting action.

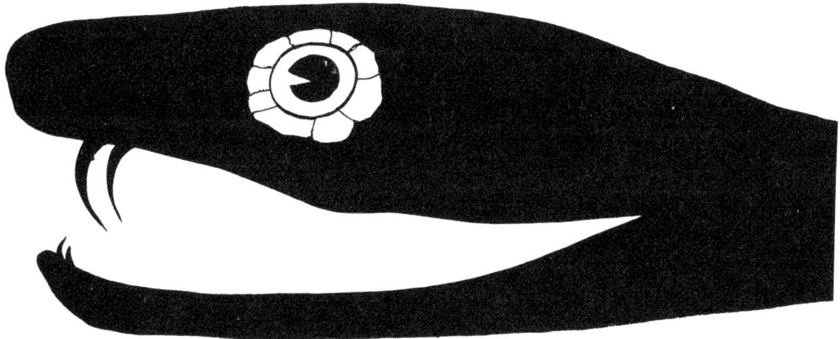

Mamba

Other members of this group are: common night adder, snouted night adder, burrowing adder, Cape cobra, Egyptian cobra, rinkhals, shield snake, South African coral snake, South West African spitting cobra, garter snake.

Boomslang—belonging to the back-fanged group. Biting is done with a chewing action, but the snake has to be able to obtain a good grip before its fangs come into operation, especially when biting on large diameter areas such as body and legs.

Boomslang

Some other members of this group are: bird snake, Natal black snake, herald snake, skaapstekers, sand snake, Cape many-spotted snake, tiger snake, beaked snake, purple-glossed snake, quill-snouted snake, mopani snake.

Snakes of South Africa can be divided into the following groups or families:

(1) Family Boidae—having remains of clawlike hind limbs either side of the vent, and solid teeth, e.g. African python.

(2) Family Colubridae—solid-toothed or grooved-fanged venomous species—this family is divided into three subfamilies:

 (*a*) Colubrinae—solid-toothed fangless snakes, e.g. house snakes.

 (*b*) Boiginae—enlarged, grooved fangs under the eye—venomous but most are no danger to man, e.g. sand snakes. The exceptions, namely, boomslang, vine snake and Natal black snake are dangerous to man.

 (*c*) Dasypeltinae—fangless snakes such as the egg-eaters.

(3) Family Elapidae—snakes with short, tubular fangs under the snout on a relatively immobile jawbone, e.g. mambas, cobras—all dangerous to man.

(4) Family Viperidae—snakes with long, canaliculated fangs on a highly mobile jawbone. All dangerous to man, e.g. puff-adder.

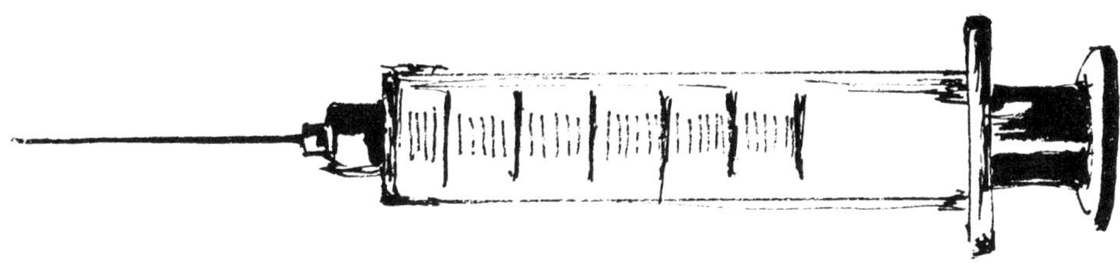

The Hydrological Cycle

(in cubic kilometres per day)

Earth's limited water supply is constantly circulating. Oceans are the principal reservoir for the earth's water. Though it is in continuous circulation, the finite supply of fresh water limits the number of people the earth can support. In order to avoid a crisis in the future, the big challenge is to make more effective use of the 100 kilometres of water which daily runs off into the ocean. Pollution must be cut so water can be used and re-used.

Less than 1 per cent of the earth's water is usable.

More than 99 per cent of the world's water is in the salty oceans, or tied up in glaciers and the polar ice-caps.

Cropland per person is dropping sharply. Food production cannot keep pace with the earth's population, which has doubled twice in the last two centuries.

We farm only 2 per cent of earth's surface. The earth's surface is 78 per cent water or ice-covered land. Virgin forest covers one-third of all land areas; crops less than 10 per cent.

PARADOXES AND PRIORITIES

As man desperately tries to raise more food to avoid starvation, he also endangers his planet. This seeming contradiction is not a question of right or wrong, but a matter of priority.

MINERALS

Fossil fuels create energy.

Energy runs industry.

Industry produces a higher standard of living.

All the world aspires to a higher standard of living. There are not enough mineral resources to support 3,7 billion people at a standard of living equal to Western Europe—at least not without important new breakthroughs in technology.

Fossil fuels now provide 96 per cent of all energy. Because they are being burned up at such an accelerated pace, experts predict that oil and natural gas will run dry within a century at present consumption rates.

Energy use is skyrocketing.

World energy use is 30 times higher than a century ago.

Expansion has been uneven.

A North American uses three times as much energy as a Western European; 30 times as much as an Asian or African.

Air pollution is an increasing hazard to health. Deaths directly traceable to air pollution are rising sharply in urban areas. Air pollution knowns no boundaries.

Air Pollution

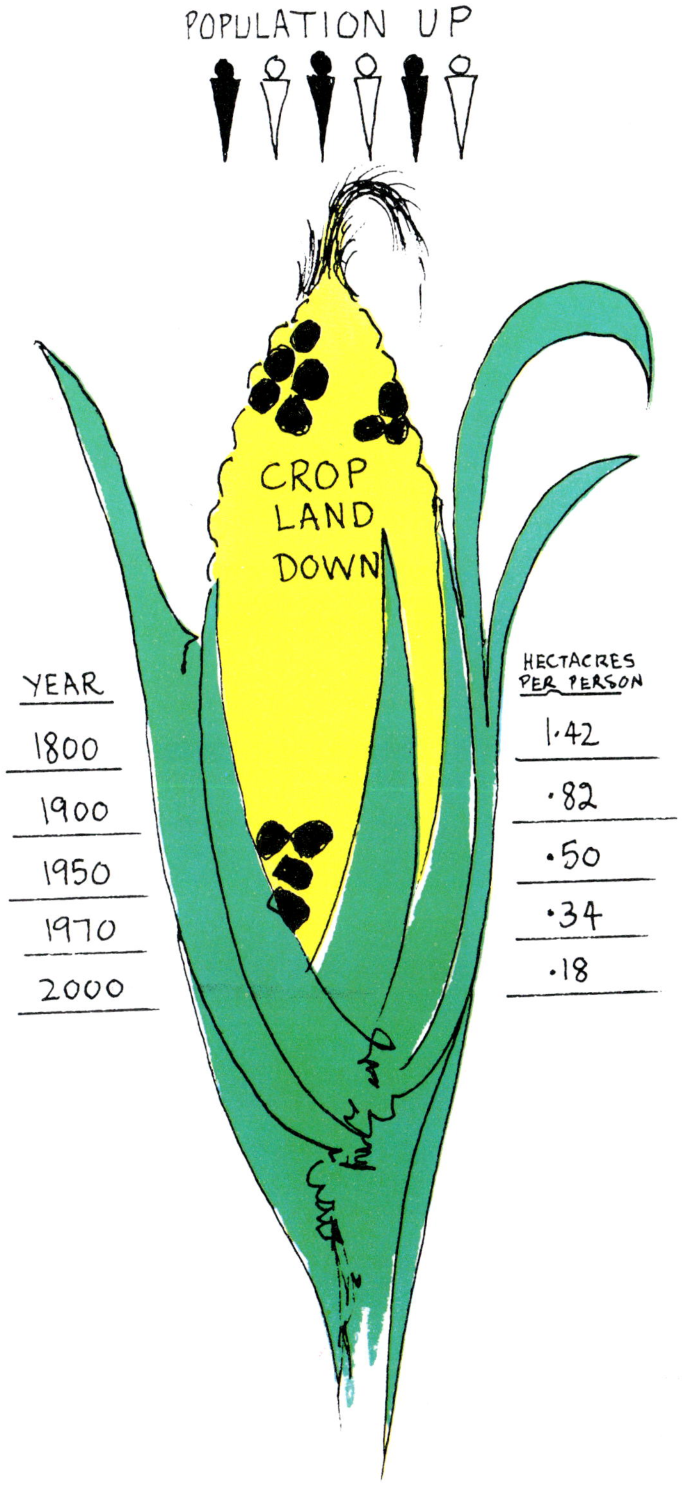

57. Food production

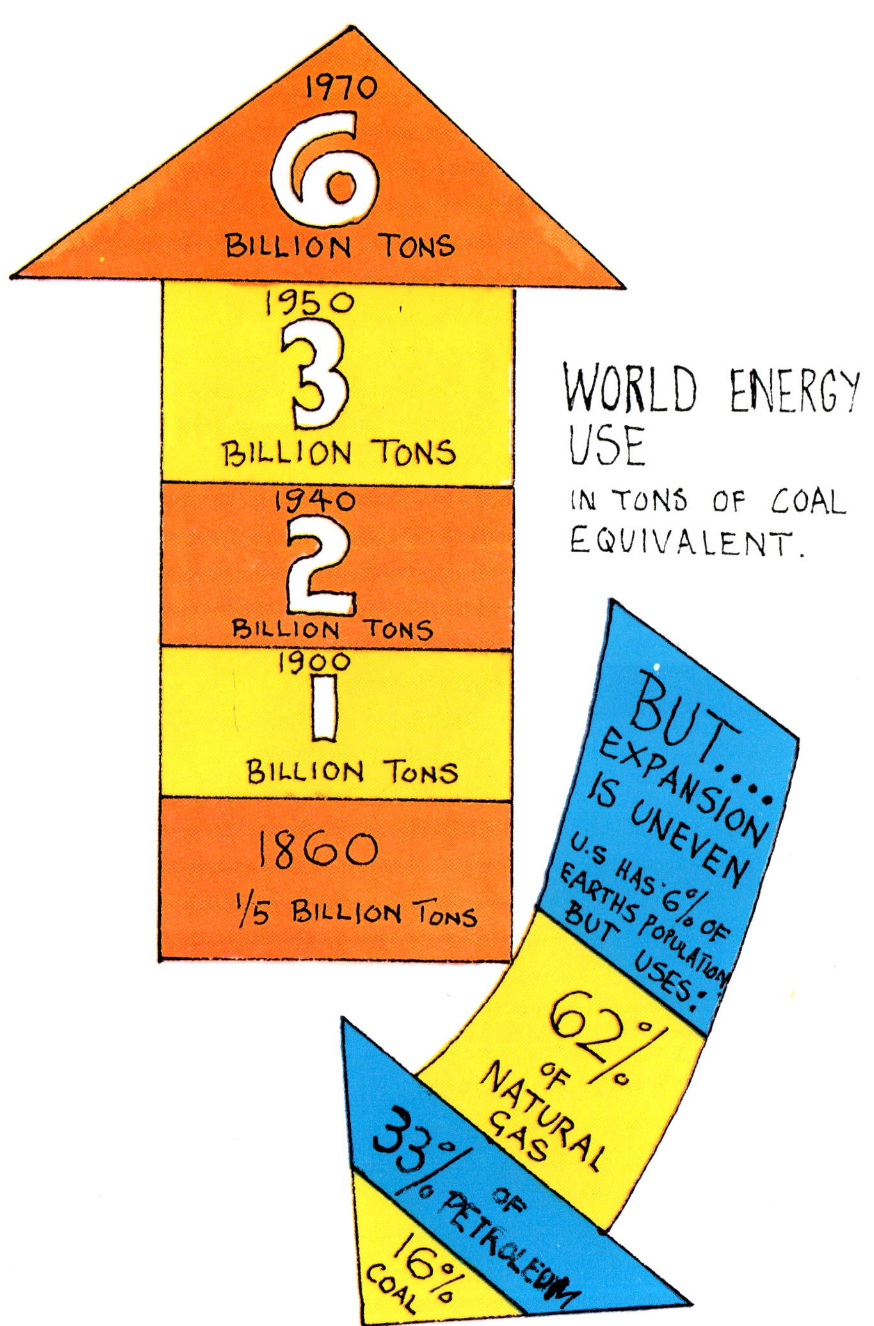

58. World energy use

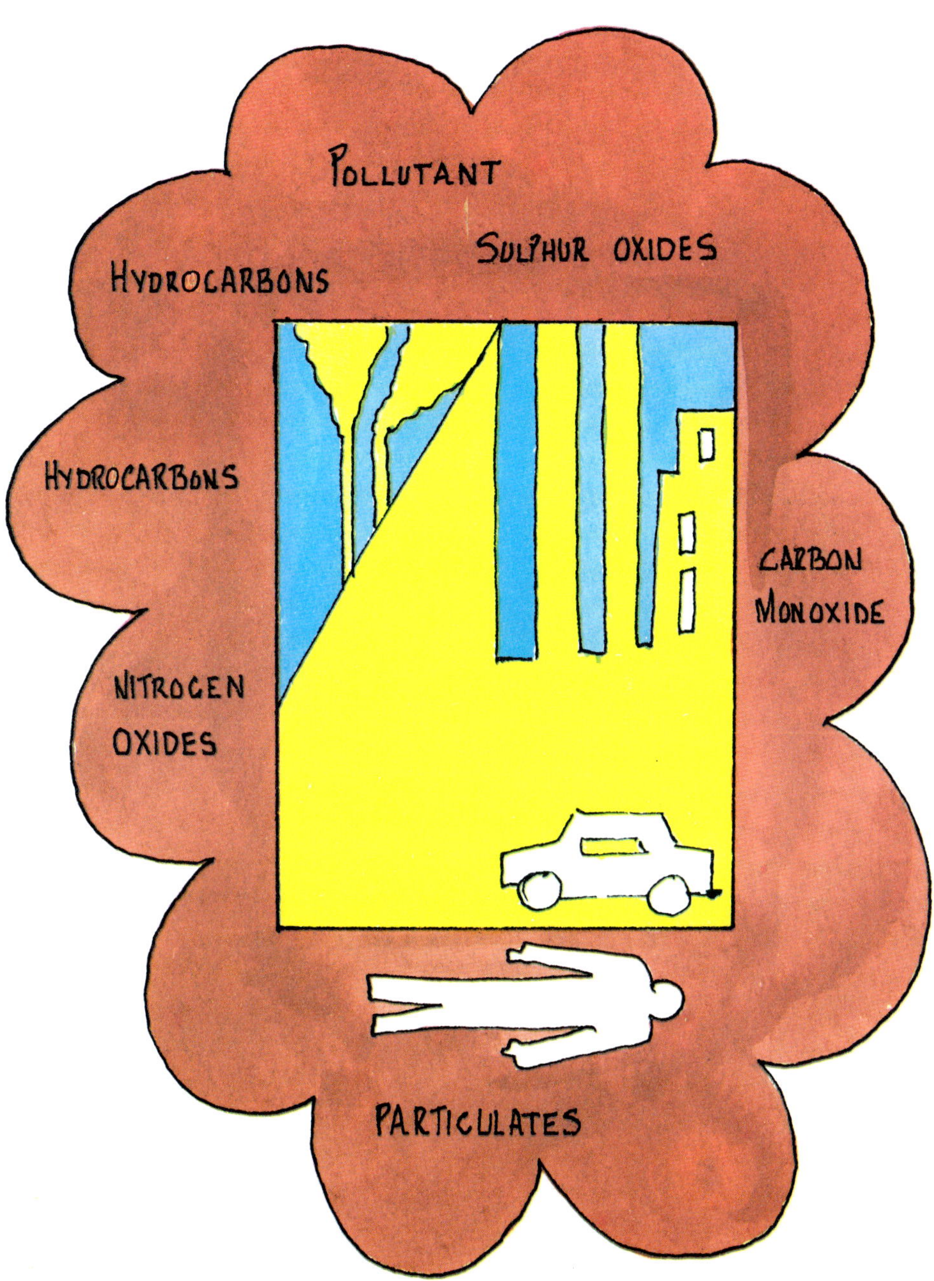

59. Air pollution

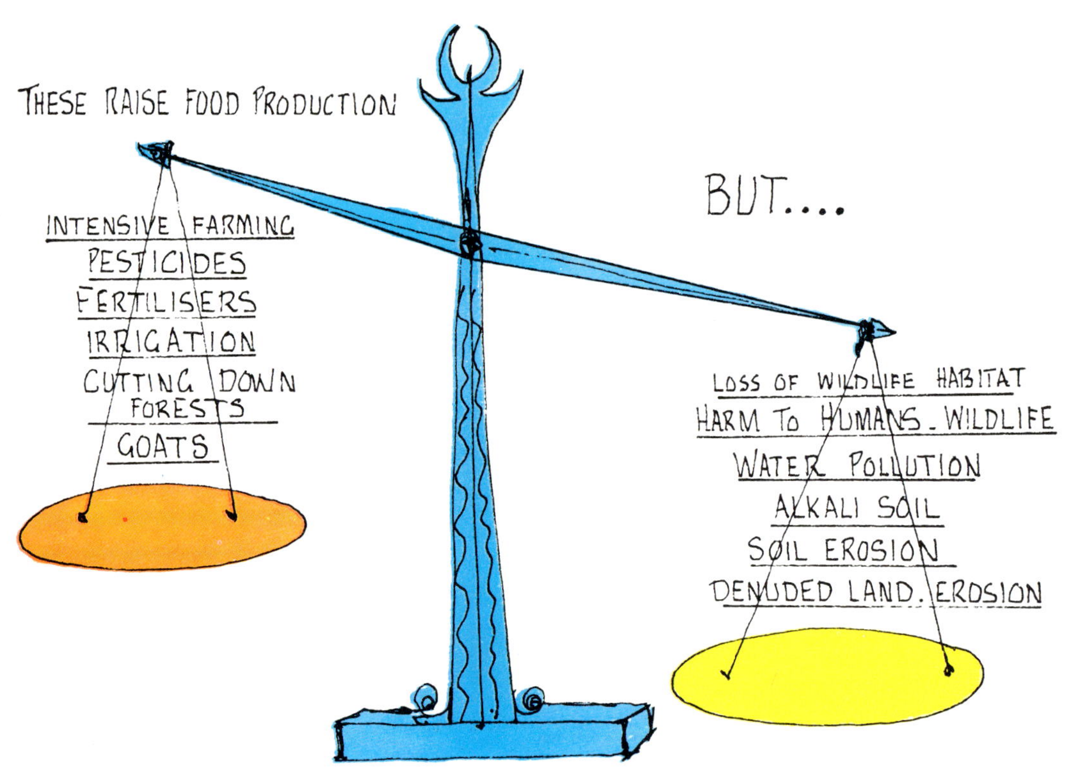

60. Paradoxes and priorities

Edible Plants of the Bushveld

Should one ever need the foods of the bushveld for basic survival, one would have to know as much as possible about poisonous and non-poisonous, edible and inedible plants.

The section which follows on edible plants of the bushveld is for the naturalist's interest. Should one be walking through the wilds, one might be keen to sample a few natural and non-poisonous berries, fruits or roots. Edible fruits have little food value, although they are refreshing and thirst-quenching. Frequently they are available only towards the end of the rainy season.

The edible roots and leaves of the more common African plants provide a substantial vegetable diet. They are available throughout most seasons and have a reasonably high nutritional content.

The *Bidens pilosa* or blackjack is a common annual weed with leaves divided into three leaflets 40 mm in length, stalked and spearhead in shape. The flowers have a yellow centre with occasional white rays. When cooked, the leaves of this weed are a good substitute for spinach.

The *Hyphaene ventricosa*, known as the gingerbread palm, fan palm or the vegetable ivory, is a slender palm with apical buds or shoots providing a useful source of food. A potent wine is sometimes distilled from the tapped sap.

The *Boscia albitrunca*, or witgatboom, is a small tree found in the southern lowveld. Its distinctive whitish trunk makes it easily recognizable at a distance. The leaves, 25 mm long, have a leathery texture, are grey-green in colour, oblong in shape and packed tightly at the end of the branches. The leaf stalks are short, the flowers tiny and the fruits are round, resembling small cherries. The fruits are edible, but have a sickly taste. The roots can be pounded and made into a type of porridge, or roasted and used as a coffee substitute. Syrup can be made if the pounded roots are boiled long enough.

The *Commiphora marlothii*, or paperbark tree, has a distinctive bark which flakes off in large, light yellow sheets, leaving a smooth, green under-bark. The roots are quite palatable in their raw state.

The *Commiphora mollis* or imiNyela is a small, attractive, leafy tree with a round, thickly-branched crown found in the lowveld, in hotter and drier types of woodland, and in river valleys at lower altitudes. It has a smooth, brown or dark grey bark mottled with circular patterns of a lighter colour, with leaves clustered together on short branchlets. The roots can be eaten raw or cooked.

SOME EDIBLE PLANTS OF THE BUSHVELD

This list is not complete. Care must be taken to ensure that identification of a plant is positive before any part of it is eaten. The scientific English, Afrikaans and Zulu names are given in that order, as well as the part of the plant that is eaten. The parts of the plant are indicated as follows:

F—fruit	S—seed
L—leaves	Rw—raw
R—root	C—cooked

Annona senegalensis, FRw.
Antidesma venosum, Tassle berry, hangbessie, isiBangamlotha, FRw.
Bauhinia esculenta, Gemsbok beans, gemsbokboontjies, SRw. R boiled for long time makes syrup. R pounded to make porridge.
Bequaertiodendron natalense, Natal sweet plum, Natalse soetpruim, umThongwane, FRw.
Berchemia discolor, FRw.
Bequaertiodendron magalismontanum, FRw.
Bidens pilosa, Blackjack, knapsekerel, uQadolo, LC.
Boscia albitrunca, Shepherd's tree, witstamboom, witgatboom, FRw. R roasted as coffee substitute or used as preservative.
Bridelia micrantha, Coast goldleaf, mitserie, iNhlalamawabayi, FRw.
Canthium mundianum, FRw.
Canthium spinosum, FRw.
Canthium ventosum (*inerme*), Turkey berry, kalkoenbessie, umVuthwamini, FRw.
Carissa bispinosa, Num-num, umVusankunzi, FRw.
Carissa dulus, FRw.
Carissa macrocorpa, FRw.
Carissa grandiflora, umThungulu, FRw.
Cassine aethiopica, Kooboo berry, koeboebessie, umNgayi, FRw.
Cephlanthus natalensis, FRw.
Chrysophyllum viridefolium, Natal plum, FRw.
Citrullus vulgaris, Melon, iBhece, FC.
Colocasia antiquorum, iDumbe, RC.
Commiphora marlothii, Paperbark tree, water from roots, RRw.
Commiphora mollis, RRw, R'C.
Commiphora mossambicensis, Paperbark tree, RRw, RC.
Cordyla africana, FRw.
Cussonia spicata, Cabbage tree, sambreelboom, umSenge. Underground roots chewed (like sugar-cane).
Dialium schlecteri, Sherbet tree, umThiba, FRw.
Dienbollia oblongifolia, FRw.

Diospyros lycoides, umNgandane, FRw (*must* be ripe or will give mild poisoning).
Diospyros mespiliformis, Rhodesian ebody, jakhalsbessie, FRw.
Dovyalis caffra, Wild apricot, wilde appelkoos, iQokolo, FRw.
Dovyalis rhamnoides, UKhamginqi. FRw.
Dovyalis rotunifolia, FRw.
Ehretia rigida, peumekaarbos, umKlele, FRw.
Eleusine caracena, Goode grass, osgras, uPhoko, SC.
Eragrostis curvula, SC.
Erythrina humeara, RRw for water.
Euclea divinorum, Gwarri, umBgopanyamazane, FRw.
Euclea shimperi, Straight-leafed gwarri, bosgwarri, iDungamuzi, FRw.
Euclea undulata, Small-leafed gwarri, kleinblaar gwarri, umShekisane, FRw.
Ficus capensis, Cape fig, Kaapse wildevy, umKhiwane, FRw.
Ficus hippopotami, Swamp fig, Moeras wildevy, iNkokhokbo, FRw.
Ficus ingens, Red leaf rock fig, rooiblaarwildevy, FRw.
Ficus sycomorus, Sycomore fig, rivier wildevy, iNcongo, FRw.
Flacourtia indica, FRw.
Garcinia livingstoni, Mangosteen, uGobandhlovu, FRw.
Gardenia neuberia, FRw.
Grewia occidentalis, Currant bush, kruisbessie, iKholo, FRw.
Halleria lucida, Wild fuchsia, FRw.
Harpephyllum caffrum, Kaffir plum, kaffirpruim, umGwenya, FRw.
Hyphaene crinita, Vegetable ivory, lala palm, iLala, Sap, FRw. S green, young leaves at ground level.
Hyphaene ventricosa, Raffia pal., FRw.
Hypoxis agenta, Sc.
Ipomoea batatas, Sweet potato, uBgatatha, RC.
Ipomoea ommanneyi, uBhoqo, RRw. or RC.
Jatropha variifolia, umDumbula, RC.
Kraussia floribunda, isiKuphankobe, FRw.
Lagenaria siceraria, Calabash, iselwa, FC.
Landolphia kirkii, Ground peach, grond-perske, imBungu, FRw.
Lantana rugosa, uBukhwebezane, FRw.
Lechnera rosea, L boiled to make tea.
Manilkara mochisia, FRw.
Mimusops caffra, Coast red milkwood, umThunzi, FRw.
Mimusops obovata, Bush red milkwood, amaSethole-abomvu, FRw.
Momordica foetida, iNtshungu, iNshublaba, LC.
Nymphaea spp. Water-lily, waterlelie, iZibu, underwater stem edible.
Oncoba spinosa, FRw.
Opantia megacantha, Prickly pear, FRw. FC (konfyt)
Pachycarpus appendiculatus, iShongwe, LC.
Pachystignia macrocalyx, FRw.
Pappea capensis, Wild plum, wildepruim, umQoqo, FRw.
Parinari curatellifolia, Cork tree, hissing tree, boomgrysappel, mobola plum, mBula, FRw.
Parinari mobola, Mobola plum, boomgrysappel, amaBuye, FRw.
Pennisetum typhoides, Napier grass, uNyawothi, SC.

Sclerocarya caffra

Parinari curatellifolia

Phaseolus arueus, umNgomeni, SC.
Phoenix reclinata, Wild date palm, iSundu, FRw.
Phyllogeiton discolor, Bird plum, umBenduza, FRw.
Phyllogeiton zeyheri, Red ivory, rooivoor, umNcaka, FRw.
Physalis peruviana, Wild gooseberry, umGungqumu, FRw.
Rhoicissus tomentosa, Wild grape, wildedruif, FRw.
Rhus chirindensis, Karee, Nhlokoshiyane, FRw.
Rhus legatii, Karee, Nhlokoshiyane, FRw.
Rhus prinoides, Karee, Nhlokoshiyane, FRw.
Rubus rigidus, Ijikijolo, FRw.
Salvadora australis, FRw.
Sarcostemma viminale, Melktou, iNgotsha, FRw and LRw.
Scutia myrtina, FRw.
Sclerocarya caffra, Maroela, umGanu, FRw, SRw.
Scolopia mundii, Mountain saffron, iliDungamuzi-lehlathi, FRw.
Scolopia zeyheri, Thorn pear, umHlambahlale, FRw.
Sideroxylon inerme, White milkwood, wit melkhoud, amaKhwelafingqane, FRw.
Sonchus cleraceus, Common sow thistle, gewone sydissel, iKalbeklabe, LC.
Sorghum dochna, amaBele, Sc.
Stychnos madagascariensis, Monkey orange, klapper, amaHlala, SRw (must be ripe).
Strychnos spinosa, Monkey orange, klapper, amaHlala, SRw.
Syzigium cordatum, Waterberry, waterbessie, umDoni, FRw.
Syzigium gerrardii, Forest waterberry, umDoni-wehlathi, FRw.
Syzigium guineense, Forest waterberry, FRw.
Tabernaemontana ventricosa, Toad tree, skurwepaddaboom, umHlambamanzi, SRw.
Trichilia emetica, Thunder tree, rooiessenhoud, mKhuklu, SRw.
Vangueria infausta, Wild medlar, wilde-mispel, umViyo, FRw.
Vigna hirta, isiKhwali, RRw.
Vigna sinensis, iMumba, SC.
Ximenia americana, Sour plum, suurpruim, umKolotshane, FRw.
Ximenia caffra, Sour plum, suurpruim, umThunduluka, FRw.
Ziziphus mucronata, Buffalo thorn, blink-blaar-wag-'n-bietjie, umPafa, FRw.

OTHER USES

Croton megalobotrys, Leaves and bark as fish poison.
Euphorbia ingens, Candelabra tree, umHlonhlo, Sap and crushed stem for fish poison. Sap boiled in water with accacia gum for bird-lime.

Trees of Southern Africa

It would be a heartbreaking thought to imagine the country stripped of its trees. One thinks of trees, rather simply, in terms of providing shade from the penetrating African sun.

Without vegetation, there would be no animal life, and in spite of the role trees play in the life of the game, trees and shrubs add a natural beauty that can never be replaced. Many trees are rare in appearance; others have long been the centre of intricate tribal customs.

The effect of rainfall on the soil determines whether one area of bushveld should have a predominance of baobabs and fever trees, while another be covered in ilala palms and mopane forests.

An interesting species is the Leguminosae family comprising the well-known acacia tree. The Leguminosae family contains over 10 000 species and the bushveld of southern Africa supports many of them. Several species, such as certain thorn trees (acacias), the mopane and the sekelbos form dense and extensive communities almost to the exclusion of others.

The observation of trees is a rewarding pastime and one soon learns the trees characterized by strong, aromatic wood, yellow-stained barks, dense thorns or thickly leafed branches, and the animals which depend on these trees for their food.

THE BUFFALO THORN TREE

(*Ziziphus mucronata*) (Blink-blaar-wag-'n-bietjie)

(Zulu—UmLahlankosi) (Xhosa—UmPhafa)

This is probably the most important or most useful tree in the bushveld. Its leaves and stems form the staple diet of the black rhino Many browsing animals also eat the leaves Its berries, when ripe, are eaten by birds, and in Damaraland the African people make a coffee from the berries.

The leaves, if boiled in water, make a palatable spinach.

The bark of this tree is used medicinally by the African people. The bark is boiled in water and the infusion is then drunk to settle stomach ailments. The boiled-up bark is used as a poultice for drawing out festering sores.

The roots of the tree are used in much the same way, except that once boiled up the mashed fibres are laid on the chest of a person suffering from chest complaints. This is reputed to draw the diseased areas clean.

The Zulu name, Lhasha Nkosi (meaning bury the chief), is derived from an old religious custom of theirs. The Zulu holds his ndaba or discussions of importance in his cattle-kraal. It is believed that it is here that the spirits of his forefathers gather, and that he can obtain guidance directly from these spirits. If the chief or head of a kraal dies, he is usually buried in the cattle-kraal, and on the first evening his grave is covered with fresh branches of the buffalo thorn tree. The cattle, which, incidentally, are browsers as well as grazers, are brought in that evening and will immediately eat the leaves of the branches laid on the grave. It is believed that now they are imbibing the spirit of their previous owner and will prosper and be easy to control.

Ziziphus mucronata

THE UMTHAMBOTHI TREE

(Jumping-bean Tree) (Tambo(o)tie)

(*Spirostachys africana*)

This tree also is a resident of the dry South African bushveld and it is well favoured as a food by the black rhino.

Porcupine, too, enjoy the underbark, and there are very few Umthambothi trees whose trunks have not been badly scarred by them. The animal seems to realize that if the tree is ring-barked it would die and his supply would diminish. Hence he always seems to leave a strip of bark so that the tree can survive.

When freshly uncovered, the newly exposed sap attracts many insects, which are in turn devoured by insect-eating birds, who are in turn eaten by birds of prey or the smaller wildcats.

The dead logs burn well in a fire, and many a story has been told round an Umthambothi camp-fire made up of old logs. The wood is hard and thus burns slowly, but the resin is also oily and inflammable and thus burns easily and well. A medium-sized log could last practically the whole night.

Beautiful furniture has also been made out of its wood and needs no furniture oil to polish it. The scent of the wood is very aromatic and pleasant and seems to linger on for ever.

The seed-pods, or beans of this tree, are often pierced during the month of November by a fruit-piercing moth, which then lays its eggs inside the seed. The young grubs then hatch out and jump around in the seed, causing the seed to fall off the tree and jump around on the ground. To anyone not used to this strange sight, it might appear as a terrible illusion to see thousands of these beans jumping on the ground. The sap of the young twigs of this tree can be extremely dangerous and painful if it enters the eyes, and should be avoided at all costs. The Africans use the sap of the *Euphorbia terychati*, heated up into a paste, as an antidote. This sap itself, on entering the eyes, can cause blindness.

SOME IMPORTANT TREES OF SOUTH AFRICA

By James Clarke the *Star*

UMBRELLA THORN (*Acacia tortilis*). Also called haak-en-steek because it has two types of thorns—hooks and straight ones. Its pods are corkscrew-shaped. This is one of the most common acacias or wag-'n-bietjie trees in the bushveld and lowveld. It is found right up to

61. Edible fruit of the bushveld species of strychnos

61

62. Without vegetation there would be no animal life

62

63

63. Collecting plant specimens is a rewarding activity

64. Environmental study areas. A spiritual impact and a lesson in the fundamentals of ecology

65. Western shore of Lake St. Lucia

66. On trail . . . schoolboys viewing the wilderness of Umfolozi Game Reserve

Ethiopia. The tree is a good indicator of deep, sweet soil and sweet grass and, therefore, a sign of the presence of winter grazing. It will grow in highveld gardens if good alkaline soil is provided.

Spirostachys africana

Mopane (*Colophospermum mopane*). Also called the butterfly tree because of the shape of its leaves. On hot days, the leaves fold (like a butterfly's wings) and present their edges to the sun so that one finds little shade in a mopane forest. The leaves are, of course, preserving moisture. The mopane (or Rhodesian ironwood) grows well in hot, brackish, sandy soil in the northern Transvaal and in Rhodesia northwards. Its leaves are vital fodder, especially in winter. Elephants are extremely fond of the mopane and its fruits.

Colophospermum mopane

Cycad (Genus: Encephalartos). The cycad or breadtree is among the world's most primitive plants. It descends from a group which dominated the earth 150 million years ago and has changed little since dinosaurs fed on it. The pineapple-like fruits are found in male and female plants, the male being distinguished by pollen clinging to it. These plants probably survive to be 500 years old and their toughness was displayed a few years ago when one, buried by a new road, burst through the tarmac into life. There are 27 different kinds in South Africa.

Red Milkwood (*Mimusops caffra*). This familiar coastal tree belongs to the family Sapotaceae—the stamvrug family. An American version produced the basic ingredient for chewing-gum. The red milkwood or moepel is an evergreen whose marble-sized fruit is the staple diet of many birds and coastal monkeys. The tree is found all along the eastern seaboard of South Africa and its importance as a dune stabilizer has protected it from over-exploitation by boat-builders, who favour its wood. The Post Office tree at Mossel Bay is a close relative.

Baobab (*Adansonia digitata*) or the kremetartboom, cream-of-tartar tree, monkey bread, and so on. This magnificent tree rarely grows more than 14 m in height, but many measure anything up to 40 m in girth. It is a tree of the hot, low rainfall, frost-free regions below 1 000 m. Its spongy, fibrous wood is of little use, although Africans use the bark fibres. The bark regenerates. Baobabs—the really big ones—may be 1 000 years old. A specimen with a diameter of 5 m was carbon-dated at 1 010 years.

Stinkwood (*Ocotea bullata*). The black stinkwood trunk, when young, is a subtle arrangement of exquisite colours. As this forest tree ages, its bark darkens. The heartwood itself is beautiful and lustrous, and its popularity in the eighteenth century caused the destruction of most of the Cape's forests. In fact, the Cape actually imported stinkwood from South America last century. In Zululand, the Zulu destroyed many grand old trees because the bark was used to stop headaches, and in the Transvaal, the early gold industry used the last of the few for mine-props.

Yellow-wood (*Podocarpus falcutus*). The massive yellow-woods are the tallest of our indigenous trees and are found in several parts of the Transvaal, Natal and the Cape. The 'Big Tree' of the Knysna Forest, which is the tallest native tree in South Africa, is well over 42 m high. It probably predates the Christian era. The early gold-mining industry cut out the Transvaal's big yellow-woods—even Johannesburg's tramline sleepers were made of yellow-wood. The tree can be fairly fast-growing—one in a Pretoria garden grew 15 m in 15 years.

Ocotea bullata

Kiepersol (*Cussonia spicata*) is also called the cabbage tree. It grows to 10 m high and its sometimes blue-grey leaves resemble cabbages in texture. They make excellent fodder and Eastern Province farmers believe they are superior to lucerne. (Analysis has shown it is not quite as good.) It has been suggested that the tree could be grown widely as a drought-resistant fodder-bank. Its club-like, sappy root is sometimes dug up by desperate farmers and fed to cattle dying of thirst. The root is common in the Transvaal's rocky hills and pocket forests.

Soetdoringboom (*Acacia Karroo*)—the tree of the folk-song 'Sarie Marais' and South Africa's most common native tree. Also called the mimosa, sweet-thorn, doringboom, etc. Perhaps, historically, it is the most useful tree in the country. The *Acacia karroo* usually indicates moist subsoil and often indicates an 'oasis' in the Karoo or in the Kalahari. It has a sweet aroma when in bloom, and the thorns, usually confined to the lower branches, protect it from grazers.

PROTEA (*Protea caffra*) or the suikerbos or sugarbush. A small, gnarled, untidy tree with magnificent blooms which exudes a sweet-tasting nectar; hence 'sugarbush'. Mostly, proteas are strictly protected and they are unique to South Africa. The name protea comes from Proteus, the sea god who took on many forms. So does the protea, of which there are many species. Its leathery leaves are reminiscent of the leaves of certain prehistoric trees which once dominated the earth.

ALOE (*Aloe ferox*). There are well over 100 species of aloes in South Africa, many of them discovered only this century. This one yields 'bitter aloes', often used medicinally. The name 'ferox' means fierce—obviously because of this plant's thorny defence, a common characteristic among aloes. These strange succulents belong to the lily family, are unrelated to cacti, and are mostly found in semi-arid areas. They can withstand withering droughts and most are easily grown in gardens from seed. They are all fully protected in the Transvaal.

KAFFERBOOM (*Erythrina lysistemon*). There are two well-known species of kafferboom in South Africa—*Erythrina caffra*, which is found mostly in the Cape and which has short and broad flowers, and its Transvaal relation *Erythrina lysistemon*. This beautiful tree is drought-resistant and can stand a few degrees of frost, too. There is a magnificent specimen in Hendrik Verwoerd Drive, Randburg, which is every bit as impressive as those found in the eastern Transvaal. The wood was used for canoes, drinking-troughs and for wagon brake-blocks.

Animal Spoor

Tracking wild animals is a skill that represents the ultimate in bushcraft. The road to tracking game by spoor identification is a rewarding experience for the naturalist with keen eyesight and memory, an understanding of the bush and the ability to concentrate.

Bushcraft skills are best acquired by watching an experienced tracker, and, armed with a guidebook, developing one's own skills with patience and practice.

In observing animal spoor, the things a tracker should look for are footprints, the rhythm of the spoor and the length of the stride, the latter being a guide as to where the next footprint can be found. Tell-tale signs include trampled grass, disturbed stones, sticks, cracks in the soil; leaves turned, crushed or pulled off trees; broken branches and twigs; vegetation pushed aside and scratched tree bark. Animals with cloven hooves have footprints with sharp edges, for example, the antelopes.

Only a tracker with considerable knowledge can establish the age of the spoor with accuracy. Sun, wind and rain may erode the trail, but a knowledge of the local conditions can be helpful in estimating the age of the tracks. The colour and texture of trampled vegetation might provide a clue, while the position of trees and bushes relative to sun may cast shadows on the spoor during the day.

Other factors which determine tracking include whether the ground is hard or soft, stony or muddy and whether the terrain is mopane forest or open savannah.

Pressing Plant Specimens

Creating a small library of flower and leaf specimens from a nature expedition is a rewarding, fun-filled activity.

You develop a deep understanding of nature when you examine plant roots, stalks and leaves in the bush. For instant recall, a scrapbook of bushveld items revives the times spent in the veld collecting weeds, grasses, flowers and seeds.

If you collected a few leaves and seeds of the marula tree, you'd remember tasting the fruit and being told how animals can get quite drunk if they eat too many marula fruits.

The energy spent in pressing plants and creating a scrapbook makes the next trail more memorable when you are able to identify a dozen plant species which meant nothing to you on the first trail.

COLLECTING AND PRESSING HERBARIUM SPECIMENS

Specimens can be pressed on the spot, or they can be kept for as long as a day in a closed plastic bag.

A plant press consists of two frames measuring approximately 31 × 46 cm and preferably constructed from wooden slats, the slats dovetailed to give one flat surface. Sheets of newspaper cut to size can be used as blotters, and folded sheets can be used as pressing papers for delicate specimens that will be spoilt by handling during the drying process. The press is built up by alternating blotters and pressing papers, where used, and sheets of corrugated cardboard can with advantage be inserted at intervals to keep the press flat and to allow some air to circulate. The frames form the top and bottom of the press, which is then 'locked up' by means of straps drawn up tightly.

The blotters need to be changed daily for about the first three days, less frequently later, depending on the material being pressed. Fungi quickly attack wet specimens and spoil them. If pressing papers are being used, they are of course left intact with their enclosed specimens. Only the interleaving blotters are changed. The press must be left in the sun (or next to a warm stove in wet weather) to hasten drying, and the wet blotters can likewise be dried and used again. It is imperative to bring the press indoors at night, away from dew or any sort of moisture.

Select specimens that are free from insects or the evidence of insect

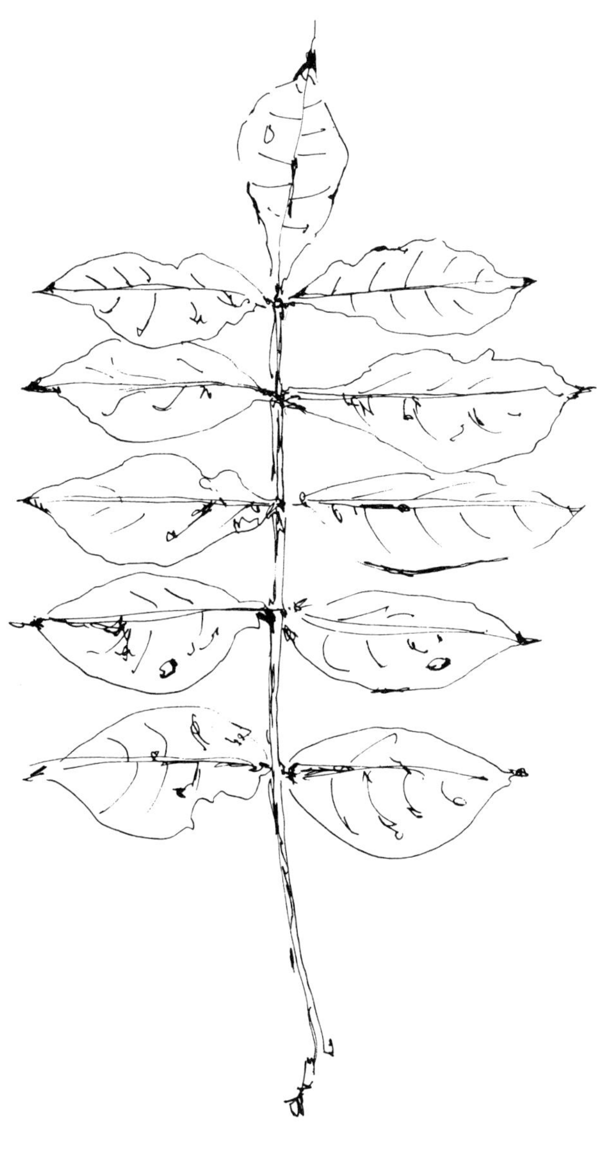

Marula

feeding, or any signs of disease. The specimens must be flowering or fruiting; sterile material is generally useless.

If the specimen is herbaceous, e.g. grassveld plants, collect the whole plant, including roots or underground organs. Plants measuring more than about 38 cm may be folded in a V or N shape, and the folds held in place, if necessary, by slips of slotted paper. *Never cut stems; always fold.*

Nyala tree

Very large leaves may have to be pressed separately, but always retain a little of the stem to show the manner in which the leaf joined the stem.

Some plants may be split in half lengthwise, and the halves pressed separately, e.g. clivia or agapanthus. Fleshy plants like these, or aloes, or anything similar, are best killed by dipping in petrol and then pressing. Delicate flowers that may stick to the blotters or pressing papers can be

Transvaal gardenia

slipped inside a fold of tissue paper or greaseproof paper. They can be peeled off carefully when quite dry.

Pads of rolled or folded newspaper can be placed around bulky specimens to aid in keeping the press flat.

Fruits can be sliced lengthwise or crosswise, whichever is more suitable, and the slices, or some of them, dried. Succulent stems, e.g. euphorbia, can be treated likewise. Scoop out as much of the pulp as possible.

Be generous with the amount of material selected for each specimen. It is often necessary to take some of the material for dissection. An herbarium sheet measures approximately 47 × 28 cm and needs to be well covered by the specimen. Collect enough for two sheets if possible.

After 24 hours, the press is opened and each specimen examined. The specimens will have become limp, and can be arranged to display all parts to best advantage. This arranging is important and makes all the difference between a good and a poor specimen. The blotters can be changed while this is being done and the press can then be strapped up for a further 24 hours.

Each specimen must be accompanied by a label, *written in pencil*, to give the name of the collector, his collector's number, if he keeps a record of all his plants, the locality (province, magisterial district, *nearest town* (this is important) or name of farm, etc.) and notes about the specimen, e.g. tree, shrub, or herb; colour of flowers and fruits; habitat, i.e. growing in grassveld, vlei, forest, etc. Date is also important.

I. *Plant Press*—Instead of slats, 12 mm masonite with holes cut in it may be used.

II. *Size of Herbarium Sheets*—26,5 cm × 40,5 cm. Obtainable from—Seligson & Clare Ltd., P.O. Box 2431, Durban. Telephone 822327. (R5 for 200 sheets.) It is called Showdrop Triplex Board 4 sheet substance!

III. *Recipe for Mercuric Chloride Solution for Poisoning* 2 per cent solution required. Weigh out 20 grams solid mercuric chloride. Drop into 1 litre rectified spirit (± 96 per cent ethyl alcohol). Add 1 crystal copper acetate. Allow to stand overnight.

IV. *Mount* with raffia, tying knots at back and cover with small squares of brown paper stuck on top.

Environmental Study Areas

STUDYING NATURE'S WAY

The most valuable means of describing the total surroundings of the environment, to youth and adults alike, is through outdoor education. To appreciate the balanced relationship of one living thing to another, an individual can absorb a great deal from studying in the wilds.

Outdoor education is increasing in popularity not only because of its motivation for youngsters to study biology and botany, for example, with new enthusiasm, but also because of the underlying advantages that outdoor education offers—leadership, discipline, a sense of responsibility and adventure in the outdoors. The result of outdoor education, through wilderness experience programmes, is an intelligent appreciation of the environment and an understanding of the relationship between soil, plants, water, air, birds, fish, other animals and minerals. An acute awareness is generated in the care of the country's natural resources.

Richard Salmon, founder and resident director of the South African Mountain Leadership School at Clarens in the Orange Free State, runs outdoor education courses to enable young South Africans to experience

The Wilderness Leadership School is a unique 'living experience'

the ways of the outdoors. The school offers youngsters an exciting and informative set of activities in an unbeatable environment. They learn bushcraft, direction finding, game-tracking, basic survival techniques, canoeing, mountaineering and game-capture methods. The Mountain Leadership's motto is 'Effort, Humility, Service' and the aim of the school is 'Leadership through an outdoor education experience'.

The Durban-based Wilderness Leadership School runs highly specialized wilderness courses which are tailored to bring home to participants the intrinsic value of wilderness and wildlife that is fast disappearing. Theirs is a course which cultivates the awareness, in schoolchildren as well as adults, of the need to conserve natural resources—our soil, water, flora, fauna, awareness of leadership qualities and the importance of self-discipline and self-reliance, as well as the urgent need for us to understand and respect the delicate balance of nature. The Wilderness Leadership School teaches the fundamentals of environmental conservation through wilderness experience. Five prominent trustees, a board of governors, are responsible for policy and management, and a staff of professional field officers and administrative personnel constitute the school.

The school was founded in 1957 by Ian Player, who realized that unless there existed a large body of well-informed, conservation-orientated leaders, irreplaceable natural resources would seriously diminish. Only by giving future leaders the stimulation of a wilderness experience would they come to appreciate the needs and laws of the natural environment in the face of advancing technology.

On course, participants are made aware, frequently for the first time, of the mismanagement and abuse of the land, as seen in silted-up riverbeds, the absence of riverine vegetation, overgrazed hill slopes and inefficient, wasteful irrigation. A visit to Lake St Lucia is the climax of the course, for there participants witness a clear example of the complete disregard that man has had for his environment. They see the dying lake as a symbol of the need for conservation.

The school's experience shows that there are very few participants who do not finish the course with a greater awareness of the problems facing the country, with a better insight into their solution, and, most important, with a greater maturity of approach to the environment.

One of the most enterprising nature conservation schools is RECCE—Rhodesian Educational Courses in Conservation and Environmental Studies. For sheer excitement and interest, there are few features in the Rhodesian educational system to match it.

The school was established in 1971 through the inspiration of some dedicated members of the Rhodesian Parks and Wildlife Service and the

Ministry of Education. The school is situated in typical Rhodesian bush-veld, close to the shore of the large dam in Mushandike National Park, between Fort Victoria and Mashaba.

Students at Mushandike

In every week of the year, this unique school takes 50 youngsters, mainly 12-year-olds, and there they learn the ways of nature. They go there to look at ants, at birds and snakes, at blades of grass. They study animals and plants, and they learn bushcraft, and how to handle a gun.

The school's teachers demonstrate to the youngsters how to manage and make wise use of nature's resources so as to ensure their continuing replenishment; the desperate need, in this increasingly technological and over-populated world, to provide sanctuaries in terms of space, wildlife and natural vegetation is emphasized, as well as the need to preserve within them a variety of natural plant and animal communities.

The area in which RECCE operates is well endowed with nature's riches—it has an abundance of wildlife and vegetation, and great areas of untouched bushveld.

Children are taken on a variety of exploration trips on land and water to study nature at first hand. Lectures and field-work are always followed by discussions. The children experience camping out in the bush, and are taught to identify grasses, birds, trees, bushes, insects, game, snakes and other wildlife.

The effect of Mushandike's field school has been so great that conservation has been initiated as a compulsory subject in all Rhodesian primary schools.

WILDERNESS EDUCATION TRUST

This was founded by graduates of the Wilderness Leadership School whose aim is to provide funds for environmental education on a subsidized basis. Courses are conducted at 'Londolozi', Sabie-Sand Wildtuin. The trust supports many organizations.

JOINT VENTURE

Joint Venture is a combined project of the Wildlife Society and the Wilderness Leadership School, designed to provide a joint environmental awareness programme.

During term-time, complete classes of primary schoolchildren, together with their teachers, are taken out on trails and then, on their return, they follow up, in the schoolroom, the living lessons of the bush.

Courses are also held over week-ends and holiday periods when not only children, but people of all age-groups, are taken on similar courses.

No bars divide man and beast on the Wilderness trail

Courses for student-teachers and teachers are conducted to enable them to evaluate the Joint Venture programme at first hand.

Field trails are undertaken and the children sleep out under the stars—experiencing the beauty of sunrises and sunsets. They experience heat, cold, scratches, getting wet and physically tired. Some love it, some don't; but, in the end, the majority come out with changed attitudes to the life around them and with the realization of the need for conservation of our wild country.

Subjects covered are energy-flow, natural resources, population explosions, game reserve management, animal adaptations, inter-relationships, associations, territorial behaviour and dynamics. A dissection of an animal is undertaken and the scholars are able to relate the anatomy of the animal to their own bodies. Field studies also cover subjects such as the make-up of soil, successions, soil erosion and geology.

It is hoped that service organizations will use the facilities offered by Joint Venture for holiday courses for children who may not otherwise have the opportunity to go out on field expeditions.

Joint venture presently operates in Natal and it is planned to expand to the other provinces.

67. Studying Nature's way—the relationship of living things to each other and to their non-living environment

Photo: E. Shelwell

68. 'Away out in the wilderness vast where the foot of man hath seldom passed'

Planning a Trail

For the nature lover, trekking the veld is a peaceful, inexpensive and rewarding way to escape the frustrations of the city life. Although the bush represents discomfort to the uninitiated, being comfortable is only a matter of a little organization.

If one is keen on planning an independent trail of one or two days near home, one would need the minimum of luggage and equipment. However, if one visits the game reserves of southern Africa, there are several travel firms recommending excellent safari tours of four days, one week and longer. One could take a safari to the Okavango Swamps, the Moremi and Chobe game reserves across the border in Botswana, or to the famous Kruger National Park in South Africa, or any one of the reserves run by the Natal Parks Board—Umfolozi, Ndumu, Mkuze, Lake St Lucia or Hluhluwe. If one is seeking a complete break from civilization, one could embark on one's own wilderness trail on foot or using a landrover, station wagon or combi vehicle. One could also consider the marvellous wilderness trails offered by the Wilderness Leadership School operating in Natal or the Wildlife Society, which arranges trails and safaris in the Cape, Transvaal, Botswana, Angola, Rhodesia and South West Africa. These trails take the form of rough camping trails, safari bus tours, mountain hikes, marine trails or wilderness trails.

Once the nature lover has experienced a few country trails, he will select his own equipment and revise his own personal checklist. The important thing is to travel light and in this day and age of dripdry fabrics this is fairly simple to achieve.

Also to be carried are matches, a flashlight, screwdriver, bottle-opener and tinned food—bully beef, baked beans, or whatever appeals to the individual's taste. Not to be forgotten is the toothbrush, or you'll be forced to improvise with a porcupine quill.

A pocketful of nuts and raisins, while walking, provides quick energy, but the best reserves of energy are the trail itself with curiosity developing round every koppie, near every tree, under an old grey stone or beneath an aged tree log.

When sleeping in the veld, set up camp well before sunset and sleep with your head towards the fire. If disturbed by snakes, spiders or animals, you will see further into the darkness with the fire behind you.

People are generally afraid of storms, snakes and other dangers. As scary as lightning may seem, it rarely strikes people. Storms are part of the outdoor experience and have a definite part to play in the environment. A snake is not aggressive unless attacked or surprised. Most slither away when they feel the ground vibrate with your footsteps. More to be feared are sunburn—you left the cream at home, uncomfortable shoes and the fact that you forgot to take a map. Most important items include camera equipment and a good pair of field binoculars.

On trail the pulse of Africa comes through your foot

An ideal equipment list, apart from your own personal additions, would include:

Waterbottles
Large ground sheet of transparent plastic or canvas
First-aid kit/snake-bite kit
knife
waterproof container of matches
tinned food
headgear
spare socks
compass
map
magnifying glass
aluminium foil
nylon cord
mosquito-netting (optional)
sandals
walking/hiking shoes (well worn in)
torch and fresh batteries
field binoculars
camera equipment
bush jacket and shirts
other waterproof clothing

VOLUNTARY ORGANIZATIONS CONCERNED WITH THE NATURAL ENVIRONMENT

The Endangered Wildlife Trust, P.O. Box 1383, Johannesburg 2000
The National Veld Trust, P.O. Box 9552, Johannesburg 2000
The S.A. Nature Union, P.O. Box 9552, Johannesburg 2000
The Vaal River Catchment Association, P.O. Box 9552, Johannesburg 2000
The Umgeni Catchment Association, P.O. Box 552, Pietermaritzburg 3200
Friends of the Earth, P.O. Box 11435, Brooklyn, Pretoria 0011
S.A. Nature Foundation, Private Bag 95, Stellenbosch 7600
Eastern Cape Wild Bird Society, P.O. Box 1305, Port Elizabeth. 6000
Rand Piscatorial Association, P.O. Box 2813, Johannesburg 2000
Association for the Protection of the Environment, 11 De Beer Street, Stellenbosch 7600
CARE Campaign, The Star, P.O. Box 1014, Johannesburg 2000
S.A. Hunters and Game Conservation Society, P.O. Box 1703, Pretoria 0001
Clean Lane Association, P.O. Box 814, Cape Town 8000
S.A. Federation of Beekeepers' Association, P.O. Box 48, Irene, Transvaal 1675
S.A. Women's Agricultural Union, 55 Albert Street, Waterkloof, Pretoria 0002
Federation of Women's Institutes, P.O. Box 153, Pietermaritzburg 3200
Outeniqualand Trust, P.O. Box 33, Sedgefield, near George 6573
United Municipal Executive of S.A., P.O. Box 1925, Pretoria 0001
Co-ordinating Council for Nature Conservation in the Cape, 29 Barmbeck Avenue, Newlands, Cape 8001
National Council of Women, 432 CTC Building, Plein Street, Cape Town 8001
Associated Scientific and Technical Societies of S.A., P.O. Box 61019, Marshalltown, Transvaal 2107

Soil Association of S.A., P.O. Box 83, Claremont, Cape 7735
S.A. Council for Conservation and Anti-Pollution, P.O. Box 100, Northway, Durban 4065
Wilderness Education Trust, 27 Glendower Avenue, Dunvegan, Edenvale 1610. Transvaal.
S.A. Pigeon Union, P.O. Box 286, Bloemfontein 9300
Institute of Landscape Architects, P.O. Box 1889, Pretoria 0001
S.A. Ornithological Society, P.O. Box 3371, Cape Town 8000
Geological Society of S.A., P.O. Box 1017, Johannesburg 2000
Eastern Cape Public Bodies, P.O. Box 134, East London 5200
The Institute for the Prevention of Water Pollution (Western Cape Branch), City Laboratories, P.O. Athlone, Cape 7760
Organic Soil Association of S.A. ,P.O. Box 47100, Parklands, Johannesburg 2121
Tree Society of S.A., P.O. Box 4116, Johannesburg 2000
S.A. Camping Club, 84 Northway, Durban North 4016
Wilderness Leadership School, P.O. Box 10536, Bellair, Natal 4006
Tvl. Branch, P.O. Box 10418, Johannesburg 2000
Zwartkops Trust, P.O. Box 1359, Port Elizabeth 6000
The Council of Boy Scouts of S.A., P.O. Box 1233, Johannesburg 2000
Operation Wild Flower, P.O. Box 31560, Braamfontein, Transvaal 2017
Institute of Parks and Recreation Administration, P.O. Box 1454, Pretoria 0001
Mountain Club of S.A. (Tvl), 5 Powers Court, Central Avenue, Atholl, Johannesburg 2001
Mountain Club of S.A. (Natal), 4 Callender Road, Westville, Durban 3630
Witwatersrand Bird Club, P.O. Box 7048, Johannesburg 2000
Johannesburg Hiking Club, P.O. Box 2254, Johannesburg 2000
Caravan Club of S.A., P.O. Box 5619, Johannesburg 2000
Geological Society of S.A., P.O. Box 1071, Johannesburg 2000
Johannesburg Council for Natural History, 6 Judith Road, Emmarentia, Johannesburg 2001
Johannesburg Council for Adult Education, 21 Empire Road, Parktown, Johannesburg 2001
Transvaal Horticultural Society, P.O. Box 7616, Johannesburg 2000
Mountain Development Association, P.O. Box 23, Paarl 7620
Voortrekkerbeweging, 37 Von Brandis Street, Marshalltown, Transvaal 2001
S.A. Agricultural Union, P.O. Box 1508, Pretoria 0001
S.A. Anglers' Union, P.O. Box 74, Pretoria 0001
Wildlife Protection and Conservation Society of S.A., P.O. Box 489, Pinetown, Natal 3600
The Girl Guide Association, 987 Park Street, Pretoria 0002
S.A. Archaeological Society, P.O. Box 31, Claremont, Cape 7735; or P.O. Box 1038, Johannesburg 2000
Round Table Associations of S.A., P.O. Box 2091, Pretoria 0001
National Association for Clean Air, P.O. Box 5777, Johannesburg 2000
Anti-Pollution Organisation, P.O. Box 113, Potchefstroom 2520
Southern African Wildlife Management, P.O. Box 411, Pretoria 0001
S.A. Co-ordinating Consumers Council, P.O. Box 3800, Pretoria 0001

S.A. National Foundation for Coastal Birds, P.O. Box 17, Rondebosch, Cape 7700

Oceanographic Research Institute, Aquarium Building, 2 West Street, Durban 4001

Caravan Club of Southern Africa, P.O. Box 50580, Randburg 2125

National Zoological Gardens of S.A., P.O. Box 754, Pretoria 0001

S.A. Forestry Association, 62 Lurgan Road, Parkview, Johannesburg 2001

The Secretary General, S.A.R.C.C.U.S., Private Bag 116, Pretoria 0001

Aloe & Succulent Society of S.A., Pretoria 0001

Natal Hunters & Game Conservation Association, 180 Longmarket Street, Pietermaritzburg, Natal 3201

University of Natal Mountain Club, c/o S.R.C., University of Natal, Pietermaritzburg and Durban 3200 and 4000

Save the Garden Route Committee, P.O. Box 2, Plettenberg Bay, Cape 6600

False Bay Conservation Society, Harmony, Kloof Road, Clifton, Cape 8001

Zoological Society of S.A., c/o University of Cape Town, Rondebosch, Cape 7700

Botanic Society of S.A., Kirstenbosch, Newlands, Cape 8000

S_2 A_3, 205 Kelvin House, 2 Holland Street, P.O. Box 61019, Marshalltown, Transvaal 2107

S.A. Underwater Union, P.O. Box 201, Rondebosch, Cape 7700

Natal Bird Club, P.O. Box 1, Snell Parade, Durban 4074

Natal Falconry Club, c/o Mr T. Oatley, Box 662, Pietermaritzburg, Natal 3200

Natal Angling Board of Control, P.O. Box 1740, Durban 4000

S.A. Anglers' Union, c/o Mr A. Jones, P.O. Box 74, Pretoria 0001

Royal Society of S.A., P.O. Box 48, Mayville, Natal 4058

Grassland Society of S.A., Mr W. Hursey, P.O. Box 3, Ermelo 2350

Museums Association of S.A., P.O. Box 61, Cape Town 8000

P.E. Aquarium, Snake Park & Oceanarium, Humewood, Port Elizabeth 6000

Durban Museum, City Hall, Smith Street, Durban 4001

Zululand Public Bodies Association, c/o Dr W. Butler, P.O. Box 95, Empangeni, Zululand 3880

Associated Chambers of Commerce of S.A., P.O. Box 694, Johannesburg 2000

International Wilderness Leadership Foundation, 150 East 42nd Street, New York, New York 10017

Durban & Coast Horticultural Society, P.O. Box 2266, Durban 4000

Rotary Clubs of Durban, P.O. Box 1953, Durban 4000

Entomological Society of S.A., P.O. Box 103, Pretoria 0001

Veld & Vlei School, Sedgefield, near George, Cape 6573

Limnological Society of S.A., c/o P.O. Box 662, Pietermaritzburg 3200

'Enviroe' (University of Natal), c/o University of Natal, Pietermaritzburg 3200

S.A. Association of Botanists, P.O. Box 471, Stellenbosch, Cape. 7600

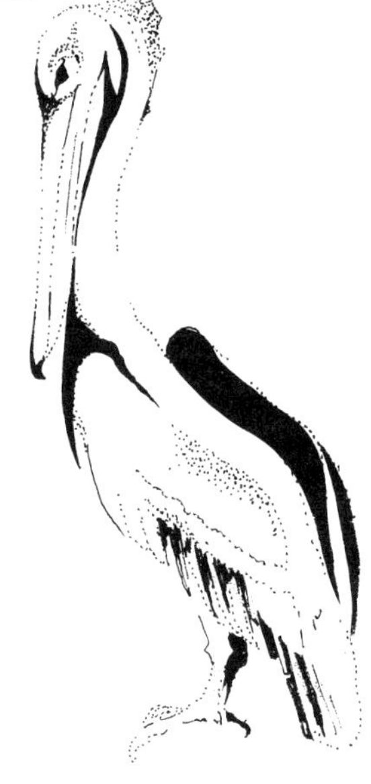

Environmental Glossary

Kudu bull and cow

Adaptation: the ability, through inherited structural or functional characteristics, that improves the survival rate of animal or plant life in a particular habitat.

Bacteria: single-celled microscopic organisms found in every habitat, ecosystem or cycle.

Carnivore: an animal that lives by eating the flesh of other animals.

Carrion: dead and decaying flesh of animals.

Commensalism: a relationship in which one partner is helped and the other is neither helped nor harmed, that is, two animals or plants existing in the same habitat.

Conservation: the wise use of the earth's natural resources that ensures their continuing availability for generations to come.

Drought: an indefinite period of time when little or no rain falls on an area.

Ecology: the study of the relationships of living things to each other and to their non-living environment.

Environment: the term which describes all external conditions such as soil, water, air and organisms, surrounding a living thing.

Erosion: the weathering of the earth's surface by water, wind, ice and other natural forces.

Grass: a plant characterized by jointed, sometimes hollow stems, two-part leaves consisting of a sheath around the stem and a long, flat blade, tiny flowers on a small spike and dry, seedlike fruits.

Grazer: an animal that feeds on grass such as a zebra or other antelope.

Habitat: the immediate surroundings of plant or animal, having everything necessary to life in a particular area.

Herbivore: an animal that thrives on plants.

Koppie: a local term for a small hill, composed of small boulders.

Mammals: the term for the group of animals including human beings, bats, cattle—all are warm-blooded, have milk-producing glands, are partially covered with hair and normally bear their young alive.

Mutualism: a relationship between two organisms either plant or animal in which both partners benefit.

Overgrazing: intensive feeding on the vegetation of an area by wild or domestic animals which causes serious and often permanent damage to the area's plant life.

Predator: an animal that lives by capturing other animals for food.

Prey: a living animal that is captured by a predator such as a lion, leopard or cheetah for food.

Pride: a family or group of lions.

Scavenger: an animal like the vulture or hyena that lives by devouring the dead remains of other animals and plants.

Shrub: a small woody plant with more than one stem rising from the ground.

Species: the term is singular or plural, and relates to a group of plants or animals with common characteristics.

Symbiosis: an association of two different organisms in a relationship that may benefit one or both partners. For example, in the case of the symbiotic relationship called parasitism, one partner (the parasite) benefits and the other partner (the host) is harmed by the association.

Territory: an animal's domain which it defends against species of its own kind or other species, or an area used by a particular animal for feeding and breeding.

Veld: a local term for open land used for grazing and other needs.

Wilderness: an area of land, whether savannah, desert or forest, where man has not settled; an area where the original natural community of plants and animals exists and survives to a balanced pattern unchanged by mechanized civilization.

Selected Reading

BIRDS

Field Guide to Birds of Southern Africa, O. P. M. Prozesky.
Birdlife in Southern Africa, K. B. Newman.
Roberts' Birds of South Africa, 3 edition, MacLachlan & Liversidge.
What Bird is that? Hazel Stokes.
Gamebirds of Southern Africa, P. A. Clancey.
African Birds of Prey, Leslie Brown.
Birds of Africa, John G. Williams & R. Fennessey.
Calburn's Birds of Southern Africa, S. Calburn.
Black Eagle Fly Free, J. A. Cottrell.
Roadside Birds of South Africa, K. B. Newman.
Garden Birds of South Africa, K. B. Newman.
The Sunbirds of Southern Africa, C. J. Skead.
Avifauna Africana, 12 Bird Studies, Norman Neumann.
Priest's Eggs of Southern African Birds.

FLORA

Wild Flowers of the Natal Drakensberg, W. R. Trouseld.
Wild Flowers of South Africa for the Garden, Una van der Spuy.
South African Shrubs and Trees for the Garden, Una van der Spuy.
South African Aloes, Barbara Jeppe.
Suid-Afrikaanse Aalwyne, Barbara Jeppe.
Trees and Shrubs of the Witwatersrand, Barbara Jeppe.

World of Nature Series:

The Flowering Wilderness (A book for the amateur and naturalist showing the variety of the world's unusual and exotic flora).

Mushrooms and Toadstools, Orbis Leisure Library Series.
The Proteaceae of South Africa, F. Rousseau.
Wild Flowers of the Witwatersrand, A. Lucas & B. Pike.
Ericas in South Africa, H. A. Baker & E. G. H. Oliver.

An introduction to the South African Orchids, E. A. C. L. E. Schelpe.
Gladiolus (A revision of the South African Species), S. A. Obermeyer, G. J. Lewis & T. Barnard.
Trees of the Kruger National Park, vols 1 & 2, P. van Wyk.
What Aloe is that?, Eric Judd.
What tree is that?, Hazel Stokes.
What Protea is that?, H. B. Rycroft.
Beauty is Necessary, Joanne Pim.

Our Living Word of Nature Series:
Life of the Forest, Jack McCormack.
Life of the Jungle, Paul W. Richards.
The Secret Life of the Forest, Richard M. Ketchum.
Epiphytic Orchids of Southern Africa, E. R. Harrison.
Forest Trees of Natal, E. J. Moll.
Acacia Species of Natal, J. H. Ross.
National Tree List.
South African Erythrina, Esmé Hennessey.
Trees of Southern Africa, Eve Palmer & Norah Pitman.
The Stapelieae of South Africa, C. A. Luckhoff.
Aloes of the South African Veld, Bornman & Hardy.
Wild Flowers of the Eastern Cape Province, Batten & Bokelmann.
Aloes of Tropical Africa and Madagascar, Reynolds.
Observations on the Genus Commiphora, B. de Winter.
Namaqualand in Flower, Sima Eliovson.
Proteas for Pleasure, Sima Eliovson.
South African Wild Flowers for the Garden, Sima Eliovson.

AGE 8–12

Macdonald Junior Reference Library:
Timber.
Plants without Flowers.
Wild Flowers.
Trees.
Flowers of the Garden.
Flowering Plants.

AGE 9 UPWARDS

Lees- en Leerreeks:
Hoe Plante Lewe.

INSECTS

Field Guide to Butterflies of Africa, Williams.
Butterflies and Moths, Orbis Leisure Library.
Insects, Orbis Leisure Library.
Life History of the South African Lycaenid Butterflies, G. C. Clarke & C. G. C. Dickson.
Insects of the World, Walter Linsenmaier.
Spiders of Southern Africa, J. H. Yates.

REPTILES

Field Guide to Snakes of Southern Africa, V. F. M. Fitzsimons.
What snake is that?, John Visser.
African Snake Stories, John Snook.
South African Frogs, V. A. Wager.
Crocodiles, C. A. W. Guggisberg.
Snakes of Africa, R. M. Isemonger.
Poisonous Snakes, John Visser.

Age 8–12

Macdonald Junior Reference Library:
Amphibians and Reptiles.

Follett Beginning Science Series:
Turtles.
The Reptiles.
Snakes.
Frogs and Toads.

Miscellaneous

Guide to Botswana.
Guide to Swaziland.
Guide to National Parks, Nature and Game Reserves of Southern Africa.
Namib, Alice Mertens.
Kruger Park Guide, Anthony Dawes.
Land Between Two Deserts, H. Zur Strassen.
South West Africa, Alice Mertens.
The Kruger National Park (*Kruger Park, 70 years*), R. S. Labuschagne.
Die Kruger Wildtuin (*Kruger Wildtuin, 70 jaar*), R. S. Labuschagne.
Memories of A Game Ranger, Harry Wolhuter.
Men, Rivers and Canoes, Ian Player.
Listen to the Wild, Susanne Hart.
Portraits of Game and Wild Animals of South Africa, W. Cornwallis Harris.
Vanishing Africa, Riccardi.
Tree where Man was born, Mattiessen & Parker.
Focus on Fauna, James Clarke & John Pitts.
The Hunted Ones, John Sinclair.
Zambezi South, John Sinclair.
Twilight of the Wild, John Sinclair.
Life Nature Series:
The Land and Wildlife of Africa.
The Land and Wildlife of Eurasia.
The Land and Wildlife of America.
Ecology.

MAMMALS

Field Guide to Larger Mammals of Africa, Dorst & Dandelot.
Gara Yaka's Domain, D. Varaday.

In the Shadow of Man, Jane Goodall.
Innocent Killers, Jane Goodall.
What antelope is that?, Paul Rose.
Game Trails of South West Africa, H. W. Zur Strassen.
Big Game and Other Mammals, Paul Rose.
Pippa's Challenge, Joy Adamson.
The Spotted Sphinx, Joy Adamson.
The Elephants of Knysna, Nick Carter.
The Life of Hippos, Bradley Smith.

Life Nature Series:
  Mammals.
Cats of Africa, John Dominis & Maitland Edey.
Born Free, Joy Adamson (Paperback).
Living Free, Joy Adamson (Paperback).
Shamba's Raiders, Bruce Kinloch.
White Rhino Saga, Ian Player.
Serengeti Kingdom of Predators, George Schaller.
The Flesh-eaters, W. T. Miller.
Mammals of Botswana, R. Smithers.
The Life and Death of Whales, Robert Burton.
Solo. The story of an African Wild Dog puppy and her pack, Hugo van Lawick.

Age 8–12

MacDonald Junior Reference Library:
  Mammals.

Follett Beginning Science Series:
  Mammals.
  Whales.
'n Ladybird-boekie:
  Soogdiere van Afrika.

Age 9 upwards

Monkeys and Apes, Library of Knowledge.
Macdonald Introduction to Nature:
  The Life of the Meat Eaters.
Purnell's Library of Knowledge:
  Monkeys and Apes.

MISCELLANEOUS

Archaeology in Southern Africa, H. C. Woodhouse.
Art on the Rocks of Southern Africa, W. Lee & H. C. Woodhouse.
Folk Tales of Southern Africa, Compiled by Prof. Partridge.
The Drakensberg of Natal, D. Liebenberg.
Ancient Ruins and Vanished Civilisations of Southern Africa, Rodger Summers.
African Genesis, Robert Ardrey.
Social Contract, Robert Ardrey.

Territorial Imperative, Robert Ardrey.
Catch Me a Colebus, Gerald Durrell.
Plains of Cambedoo, Eve Palmer.
The Tsavo Story, Daphne Sheldrick.
Van Jagter tot Wildliefhebber, P. J. Schoeman.
Uit die dagboek van 'n Wildbewaarder, P. J. Schoeman.
Discovering South Africa, T. V. Bulpin.
Lost Trails of the Transvaal, T. V. Bulpin.
Geology of Southern Africa, Prof. E. Mountain.
An introduction to the Historical Geology of South Africa, J. F. Truswell.
Natal and the Zulu Country, T. V. Bulpin.
Wankie, Ted Davidson.
Valley of the Ironwoods, Allan Wright.
Game Ranger on Horseback, Nick Steele.
Take a horse to the Wilderness, Nick Steele.
The Ivory Trail, T. V. Bulpin.
Artists Safari, Ralph Thompson.
Artist in Africa, David Shepherd.
Joy Adamson's Africa, Joy Adamson.
The Hunter's Death, T. V. Bulpin.
Lowveld Trails, T. V. Bulpin.
Ostrich Country, Fay Goldie.
World of Nature Series:
  The Dawn of Life.

Living World of Nature Series:
  Life of the African Plains.
  Life of the Mountains.
  Life of the Far North.
  Life of the Cave.
  Life of the Desert.
High Tide, Prof. J. L. B. Smith.

MISCELLANEOUS

Rock Art of Southern Africa, C. K. Cooke.
Marine Shells of Southern Africa, D. H. Kennelly.
The World of Big Game, Penny Miller & C. T. Astley-Marberley.
African Genesis, Robert Ardrey.
Territorial Imperative, Robert Ardrey.

AGE 9 UPWARDS:

Purnell's Library of Knowledge:
  Prehistoric Animals.
Macdonald Atlas Library:
  Africa.
World of Nature Atlas, Walt Disney.
Purnell's *Discovering Nature.*
Purnell's *Concise Encyclopaedia of Nature.*

RECORDS—PLATE

12″ Long Play

Nature's Melody, Dick Reucassel & D. Adendorff.
Calls of the Bushveld, Dick Reucassel & D. Adendorff.
Klanke van die Bosveld, Dick Reucassel & Tony Pooley.
Journey Across Africa, Rowan Martin.
Reis Dwars oor Afrika, Rowan Martin.
Birdsong of Southern Africa, A. Walker, Series No. 6.
Garden Birds of South Africa, A. Walker, Series No. 7.
Bird Song of Amanzi, June Stannard.
Bird Songs of the Forest, June Stannard.
Birds of the Drakensberg, Toney Henley & Tony Pooley.
Sounds of Tongoland, Tony Pooley.
Story and Songs of Living Free and Born Free, Original Soundtrack.
Sounds of our African Heritage Wild.
Africa—Sounds from the African Bushlands.
South African Souvenir.

10″ Long Play

The Rhino Story, Jones, Brown & Smith.

7″ Extended Play

Birds of the Kruger National Park, Clem Haagner, Series No. 1.
Birds of the Kruger National Park, Clem Haagner, Series No. 2.
Wild Animals of Southern Africa, Clem Haagner, Series No. 3.
The Birds as Musicians, No. 10, Chorister and Natal Robin, Jean Claude Roché.
The Birds as Musicians, No. 11, Whitethroated and Cape Robin, Jean Claude Roché.
The Birds as Musicians, No. 12, Bokmakierie and White browed Robin, Jean Claude Roché.

TAPE CASETTE

Calls of the Bushveld, Dick Reucassel & Tony Pooley.

Bibliography

Ecological Animal Geography, W. C. Allee & K. P. Schmidt, 2 edition. John Wiley & Sons Inc., New York.

Desert Animals—Physiological Problems of Heat & Water, Kurt Schmidt-Nielsen. Oxford at the Clarendon Press, 1964.

A Guide to Marine Life on South African Shores, J. H. Day. A. A. Balkema, Cape Town, 1969.

Fundamentals of Ecology, E. P. Odum, 2 edition. W. B. Saunders Company, Philadelphia & London.

Plants and Environment, R. F. Daubenmire, 2 edition. John Wiley & Sons Inc., New York. Chapman & Hall Ltd, London.

Handbook for Farmers in South Africa, Dept of Agriculture & Forestry, 3 and enlarged edition.
Elements of Ecology, G. L. Clarke. John Wiley & Sons Inc., New York. Chapman & Hall Ltd, London.
Introduction to the Study of Animal Populations, H. G. Andrewartha. Methuen & Co. Ltd, London.
Southern Africa—A Geographical Study, vol. 1, John H. Wellington. Cambridge at the Universal Press, 1955.
Veld Types of South Africa, J. P. H. Acocks. Government Printer, Pretoria.
The Morphology of the Earth, Lester C. King. Oliver & Boyd, Edinburgh & London.
Wildlife Biology, Raymond F. Dasmann. John Wiley & Sons Inc., 1964.
Introduction to the Study of Animal Populations, H. G. Andrewartha. Butler & Tanner Ltd, Frome & London, 1961.
Fundamentals of Ecology, Eugene P. Odum in collaboration with Howard T. Odum, 2 edition. Saunders, Philadelphia, 1959.
Poisonous Snakes of Southern Africa, John Visser, Howard Timmins, Cape Town.
World E Q Index, National Wildlife Federation, Washington, U.S.A.
Custos, June 1974, P. van Wyk.
James Clarke, 'Star' Newspaper.

CLIVE WALKER